STUDENT'S SOLUTIONS MANUAL

DALE BUSKE, PH.D.

St. Cloud State University

EXCURSIONS IN MODERN MATHEMATICS

NINTH EDITION

Peter Tannenbaum

California State University–Fresno

Pearson

The author and publisher of this book have used their best efforts in preparing this book. These efforts include the development, research, and testing of the theories and programs to determine their effectiveness. The author and publisher make no warranty of any kind, expressed or implied, with regard to these programs or the documentation contained in this book. The author and publisher shall not be liable in any event for incidental or consequential damages in connection with, or arising out of, the furnishing, performance, or use of these programs.

Reproduced by Pearson from electronic files supplied by the author.

Copyright © 2018 by Pearson Education, Inc.
Publishing as Pearson, 501 Boylston Street, Boston, MA 02116.

1 17

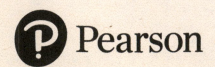

ISBN-13: 978-0-13-446913-3
ISBN-10: 0-13-446913-5

Table of Contents

Chapter 1

WALKING

1.1. Ballots and Preference Schedules

1.

Number of voters	5	3	5	3	2	3
1st choice	A	A	C	D	D	B
2nd choice	B	D	E	C	C	E
3rd choice	C	B	D	B	B	A
4th choice	D	C	A	E	A	C
5th choice	E	E	B	A	E	D

This schedule was constructed by noting, for example, that there were five ballots listing candidate C as the first preference, candidate E as the second preference, candidate D as the third preference, candidate A as the fourth preference, and candidate B as the last preference.

3. (a) $5 + 5 + 3 + 3 + 3 + 2 = 21$

   (b) 11. There are 21 votes all together. A majority is more than half of the votes, or at least 11.

   (c) Chavez. Argand has 3 last-place votes, Brandt has 5 last-place votes, Chavez has no last-place votes, Dietz has 3 last-place votes, and Epstein has $5 + 3 + 2 = 10$ last-place votes.

5.

Number of voters	37	36	24	13	5
1st choice	B	A	B	E	C
2nd choice	E	B	A	B	E
3rd choice	A	D	D	C	A
4th choice	C	C	E	A	D
5th choice	D	E	C	D	B

Here Brownstein was listed first by 37 voters. Those same 37 voters listed Easton as their second choice, Alvarez as their third choice, Clarkson as their fourth choice, and Dax as their last choice.

7.

Number of voters	14	10	8	7	4
A	2	3	1	5	3
B	1	1	2	3	2
C	5	5	5	2	4
D	4	2	4	1	5
E	3	4	3	4	1

Here 14 voters had the same preference ballot listing B as their first choice, A as their second choice, E as their third choice, D as their fourth choice, and C as their fifth and last choice.

9.

Number of voters	255	480	765
1st choice	L	C	M
2nd choice	M	M	L
3rd choice	C	L	C

$(0.17)(1500) = 255$; $(0.32)(500) = 480$; The remaining voters (51% of 1500 or 1500-255-480=765) prefer M the most, C the least, so that L is their second choice.

1.2. Plurality Method

11. **(a)** *C.* *A* has 15 first-place votes. *B* has $11 + 8 + 1 = 20$ first-place votes. *C* has 27 first-place votes. *D* has 9 first-place votes. *C* has the most first-place votes with 27 and wins the election.

    (b) *C, B, A, D.* Candidates are ranked according to the number of first-place votes they received (27, 20, 15, and 9 for *C, B, A,* and *D* respectively).

13. **(a)** *C.* *A* has 5 first-place votes. *B* has $4 + 2 = 6$ first-place votes. *C* has $6 + 2 + 2 + 2 = 12$ first-place votes. *D* has no first-place votes. *C* has the most first-place votes with 12 and wins the election.

    (b) *C, B, A, D.* Candidates are ranked according to the number of first-place votes they received (12, 6, 5, and 0 for *C, B, A,* and *D* respectively).

15. **(a)** *D.* *A* has 11% of the first-place votes. *B* has 14% of the first-place votes. *C* has 24% of the first-place votes. *D* has $23\% + 19\% + 9\% = 51\%$ of the first-place votes. *E* has no first-place votes. *D* has the largest percentage of first-place votes with 51% and wins the election.

    (b) *D, C, B, A, E.* Candidates are ranked according to the percentage of first-place votes they received (51%, 24%, 14%, 11% and 0% for *D, C, B, A,* and *E* respectively).

17. **(a)** *A.* *A* has $5 + 3 = 8$ first-place votes. *B* has 3 first-place votes. *C* has 5 first-place votes. *D* has $3 + 2 = 5$ first-place votes. *E* has no first-place votes. *A* has the most first-place votes with 8 and wins the election.

    (b) *A, C, D, B, E.* Candidates are ranked according to the number of first-place votes they received (8, 5, 5, 3, and 0 for *A, C, D, B,* and *E* respectively). Since both candidates *C* and *D* have 5 first-place votes, the tie in ranking is broken by looking at last-place votes. Since *C* has no last-place votes and *D* has 3 last-place votes, candidate *C* is ranked above candidate *D.*

19. **(a)** *A.* *A* has $5 + 3 = 8$ first-place votes. *B* has 3 first-place votes. *C* has 5 first-place votes. *D* has $3 + 2 = 5$ first-place votes. *E* has no first-place votes. *A* has the most first-place votes with 8 and wins the election. (Note: This is exactly the same as in Exercise 17(a).)

    (b) *A, C, D, B, E.* Candidates are ranked according to the number of first-place votes they received (8, 5, 5, 3, and 0 for *A, C, D, B,* and *E* respectively). Since both candidates *C* and *D* have 5 first-place votes, the tie in ranking is broken by a head-to-head comparison between the two. But candidate *C* is ranked higher than *D* on $5 + 5 + 3 = 13$ of the 21 ballots (a majority). Therefore, candidate *C* is ranked above candidate *D.*

1.3. Borda Count

21. **(a)** *A* has $4 \times 15 + 3 \times (9 + 8 + 1) + 2 \times 11 + 1 \times 27 = 163$ points.
    B has $4 \times (11 + 8 + 1) + 3 \times 15 + 2 \times (27 + 9) + 1 \times 0 = 197$ points.
    C has $4 \times 27 + 3 \times 0 + 2 \times 8 + 1 \times (15 + 11 + 9 + 1) = 160$ points.
    D has $4 \times 9 + 3 \times (27 + 11) + 2 \times (15 + 1) + 1 \times 8 = 190$ points.
    The winner is *B.*

    (b) *B, D, A, C.* Candidates are ranked according to the number of Borda points they received.

23. **(a)** *A* has $4 \times 5 + 3 \times 2 + 2 \times (6 + 2) + 1 \times (4 + 2 + 2) = 50$ points.
    B has $4 \times (4 + 2) + 3 \times (2 + 2) + 2 \times 2 + 1 \times (6 + 5) = 51$ points.
    C has $4 \times (6 + 2 + 2 + 2) + 3 \times 0 + 2 \times (5 + 4 + 2) + 1 \times 0 = 70$ points.
    D has $4 \times 0 + 3 \times (6 + 5 + 4 + 2) + 2 \times 2 + 1 \times (2 + 2) = 59$ points.
    The winner is *C.*

(b) *C, D, B, A.* Candidates are ranked according to the number of Borda points they received.

25. Here we can use a total of 100 voters for simplicity.

A has $5 \times 11 + 4 \times (24 + 23 + 19) + 3 \times (14 + 9) + 2 \times 0 + 1 \times 0 = 388$ points.

B has $5 \times 14 + 4 \times 0 + 3 \times (24 + 11) + 2 \times 23 + 1 \times (19 + 9) = 249$ points.

C has $5 \times 24 + 4 \times (14 + 11 + 9) + 3 \times 23 + 2 \times 19 + 1 \times 0 = 363$ points.

D has $5 \times (23 + 19 + 9) + 4 \times 0 + 3 \times 0 + 2 \times 14 + 1 \times (24 + 11) = 318$ points.

E has $5 \times 0 + 4 \times 0 + 3 \times 19 + 2 \times (24 + 11 + 9) + 1 \times (23 + 14) = 182$ points.

The ranking (according to Borda points) is *A, C, D, B, E.*

27. Cooper had $3 \times 49 + 2 \times 280 + 1 \times 316 = 1023$ points.

Gordon had $3 \times 37 + 2 \times 432 + 1 \times 275 = 1250$ points.

Mariota had $3 \times 788 + 2 \times 74 + 1 \times 22 = 2534$ points.

The ranking (according to Borda points) is Mariota (2534), Gordon (1250), and Cooper (1023).

29. Each ballot has $4 \times 1 + 3 \times 1 + 2 \times 1 + 1 \times 1 = 10$ points that are awarded to candidates according to the Borda count. With 110 voters, there are a total of $110 \times 10 = 1100$ Borda points. So D has $1100 - 320 - 290 - 180 = 310$ Borda points. The ranking is thus A (320), D (310), B (290), and C (180).

1.4. Plurality-with-Elimination

31. (a) *A* is the winner. Round 1:

Candidate	A	B	C	D
Number of first-place votes	15	20	27	9

D is eliminated.

Round 2: The 9 first-place votes originally going to *D* now go to *A*.

Candidate	A	B	C	D
Number of first-place votes	24	20	27	

B is eliminated.

Round 3: There are $8 + 1 = 9$ first-place votes originally going to *B* that now go to *A*. There are also 11 first-place votes going to *B* that would now go to *D*. But, since D is already eliminated, these 11 first-place votes go to *A*.

Candidate	A	B	C	D
Number of first-place votes	44		27	

Candidate *A* now has a majority of the first-place votes and is declared the winner.

(b) A complete ranking of the candidates can be found by noting in part (a) when each candidate was eliminated. Since *D* was eliminated first, it is ranked last. Since *B* was eliminated next, it is ranked next to last. The final ranking is hence *A, C, B, D.*

33. (a) *C* is the winner. Round 1:

Candidate	A	B	C	D
Number of first-place votes	5	6	12	0

Candidate *C* has a majority of the first-place votes and is declared the winner.

(b) To determine a ranking, we ignore the fact that C wins and at the end of round 1, D is the first candidate eliminated. Round 2: No first-place votes are changed.

Candidate	A	B	C	D
Number of first-place votes	5	6	12	

A is eliminated. Round 3: There are 5 first-place votes originally going to A that now go to C.

Candidate	A	B	C	D
Number of first-place votes		6	17	

The final ranking is C, B, A, D.

35. Round 1:

Candidate	A	B	C	D	E
Number of first-place votes	8	8	7	6	0

E is eliminated.
Round 2: No first-place votes change.

Candidate	A	B	C	D	E
Number of first-place votes	8	8	7	6	

D is eliminated.
Round 3: There are 4 first-place votes for D that move to candidate C and there are 2 first-place votes for D that move to candidate B.

Candidate	A	B	C	D	E
Number of first-place votes	8	10	11		

A is eliminated.
Round 4: There are 5 first-place votes for A that move to candidate B and there are 3 first-place votes for A that move to candidate C (since D and E have both been eliminated).

Candidate	A	B	C	D	E
Number of first-place votes		15	14		

B now has a majority of the first-place votes and is declared the winner. The final ranking is B, C, A, D, E.

37. (a) D is the winner. Round 1:

Candidate	A	B	C	D	E
Percentage of first-place votes	11	14	24	51	0

Candidate D has a majority of the first-place votes and is declared the winner.

(b) To determine a ranking, we ignore the fact that D wins and at the end of round 1, E is the first candidate eliminated.
Round 2: No first-place votes are changed.

Candidate	A	B	C	D	E
Percentage of first-place votes	11	14	24	51	

A is eliminated.
Round 3: The 11% of the first-place votes that went to A now go to C.

Candidate	A	B	C	D	E
Percentage of first-place votes		14	35	51	

B is eliminated.
Round 4: The 14% of the first-place votes that went to B now go to C.

Candidate	A	B	C	D	E
Percentage of first-place votes			49	51	

A complete ranking of the candidates can be found by noting when each candidate was eliminated. The final ranking is hence D, C, B, A, E.

39. Round 1:

Candidate	A	B	C	D	E
Number of first-place votes	8	8	7	6	0

Candidates *E*, *D*, and *C* are all eliminated.
Round 2: There are 4 first-place votes for *D* that go to *B* (since *C* has been eliminated). There are 2 first-place votes for *D* that go to *B*. There are 7 first-place votes for *C* that go to *A* (since both *D* and *E* are eliminated).

Candidate	A	B	C	D	E
Number of first-place votes	15	14			

A now has a majority of the first-place votes and is declared the winner.

1.5.   Pairwise Comparisons

41. (a) Candidate *D* is the winner.
  A versus *B*: $15 + 9 = 24$ votes to $27 + 11 + 8 + 1 = 47$ votes (*B* wins). *B* gets 1 point.
  A versus *C*: $15 + 11 + 9 + 8 + 1 = 44$ votes to 27 votes (*A* wins). *A* gets 1 point.
  A versus *D*: $15 + 8 + 1 = 24$ votes to $27 + 11 + 9 = 47$ votes (*D* wins). *D* gets 1 point.
  B versus *C*: $15 + 11 + 9 + 8 + 1 = 44$ votes to 27 votes (*B* wins). *B* gets 1 point.
  B versus *D*: $15 + 11 + 8 + 1 = 35$ votes to $27 + 9 = 36$ votes (*D* wins). *D* gets 1 point.
  C versus *D*: $27 + 8 = 35$ votes to $15 + 11 + 9 + 1 = 36$ votes. (*D* wins). *D* gets 1 point.
  The final tally is 1 point for *A*, 2 points for *B*, 0 points for *C*, and 3 points for *D*.

(b) A complete ranking for the candidates is found by tallying points. In this case, the final ranking is *D* (3 points), *B* (2 points), *A* (1 point), and *C* (0 points).

43. (a) Candidate *C* is the winner.
  A versus *B*: $6 + 5 = 11$ votes to $4 + 2 + 2 + 2 + 2 = 12$ votes (*B* wins). *B* gets 1 point.
  A versus *C*: $5 + 2 = 7$ votes to $6 + 4 + 2 + 2 + 2 = 16$ votes (*C* wins). *C* gets 1 point.
  A versus *D*: $5 + 2 + 2 = 9$ votes to $6 + 4 + 2 + 2 = 14$ votes (*D* wins). *D* gets 1 point.
  B versus *C*: $4 + 2 = 6$ votes to $6 + 5 + 2 + 2 + 2 = 17$ votes (*C* wins). *C* gets 1 point.
  B versus *D*: $4 + 2 + 2 + 2 = 10$ votes to $6 + 5 + 2 = 13$ votes (*D* wins). *D* gets 1 point.
  C versus *D*: $6 + 2 + 2 + 2 + 2 = 10$ votes to $5 + 4 = 9$ votes. (*C* wins). *C* gets 1 point.
  The final tally is 0 points for *A*, 1 point for *B*, 3 points for *C*, and 2 points for *D*.

(b) A complete ranking for the candidates is found by tallying points. In this case, the final ranking is *C* (3 points), *D* (2 points), *B* (1 point), and *A* (0 points).

45. Candidate *D* is the winner.
  A versus *B*: $24\% + 23\% + 19\% + 11\% + 9\% = 86\%$ of the votes to 14% of the votes (*A* wins).
  A versus *C*: $23\% + 19\% + 11\% = 53\%$ of the votes to $24\% + 14\% + 9\% = 47\%$ of the votes (*A* wins).
  A versus *D*: $24\% + 14\% + 11\% = 49\%$ of the votes to $23\% + 19\% + 9\% = 51\%$ of the votes (*D* wins).
  A versus *E*: $24\% + 23\% + 19\% + 14\% + 11\% + 9\% = 100\%$ of the votes to 0% of the votes (*A* wins).
  B versus *C*: 14% of the votes to $24\% + 23\% + 19\% + 11\% + 9\% = 86\%$ of the votes (*C* wins).
  B versus *D*: $24\% + 14\% + 11\% = 49\%$ of the votes to $23\% + 19\% + 9\% = 51\%$ of the votes (*D* wins).
  B versus *E*: $24\% + 23\% + 14\% + 11\% = 72\%$ of the votes to $19\% + 9\% = 28\%$ of the votes (*B* wins).
  C versus *D*: $24\% + 14\% + 11\% = 49\%$ of the votes to $23\% + 19\% + 9\% = 51\%$ of the votes (*D* wins).
  C versus *E*: $24\% + 23\% + 14\% + 11\% + 9\% = 81\%$ of the votes to 19% of the votes (*C* wins).
  D versus *E*: $23\% + 19\% + 14\% + 9\% = 65\%$ of the votes to $24\% + 11\% = 35\%$ of the votes (*D* wins).
  The final tally is 3 points for *A*, 1 point for *B*, 2 points for *C*, 4 points for *D*, and 0 points for *E*.

47. *A* versus *B*: $7 + 5 + 3 = 15$ votes to $8 + 4 + 2 = 14$ votes (*A* wins). *A* gets 1 point.

A versus *C*: $8 + 5 + 3 = 16$ votes to $7 + 4 + 2 = 13$ votes (*A* wins). *A* gets 1 point.

A versus *D*: $8 + 5 + 3 = 16$ votes to $7 + 4 + 2 = 13$ votes (*A* wins). *A* gets 1 point.

A versus *E*: $5 + 3 + 2 = 10$ votes to $8 + 7 + 4 = 19$ votes (*E* wins). *E* gets 1 point.

B versus *C*: $8 + 5 + 2 = 15$ votes to $7 + 4 + 3 = 14$ votes (*B* wins). *B* gets 1 point.

B versus *D*: $8 + 5 = 13$ votes to $7 + 4 + 3 + 2 = 16$ votes (*D* wins). *D* gets 1 point.

B versus *E*: $8 + 5 + 4 + 2 = 19$ votes to $7 + 3 = 10$ votes (*B* wins). *B* gets 1 point.

C versus *D*: $8 + 7 + 5 = 20$ votes to $4 + 3 + 2 = 9$ votes (*C* wins). *C* gets 1 point.

C versus *E*: $7 + 5 + 4 + 2 = 18$ votes to $8 + 3 = 11$ votes (*C* wins). *C* gets 1 point.

D versus *E*: $5 + 4 + 3 + 2 = 14$ votes to $8 + 7 = 15$ votes (*E* wins). *E* gets 1 point.

The final tally is 3 points for *A*, 2 points for *B*, 2 points for *C*, 1 point for *D*, and 2 points for *E*. Now *B*, *C*, and *E* each have 2 points. In head-to-head comparisons, *B* beats *C* and *B* beats *E* so that *B* is ranked higher than *C* and *E*. Also, *C* beats *E* so *C* is ranked higher than *E* as well. The final ranking is thus *A, B, C, E, D.*

49. **(a)** With five candidates, there are a total of $4 + 3 + 2 + 1 = 10$ pairwise comparisons. Each candidate is part of 4 of these (one against each other candidate). So, to find the number of points each candidate earns, we simply subtract the losses from 4. The 10 points are distributed as follows: *E* wins $1\frac{1}{2}$ points, *D* wins $2\frac{1}{2}$ points, *C* gets 3 points, *B* gets 2 points, and *A* gets the remaining $10 - 1\frac{1}{2} - 2\frac{1}{2} - 3 - 2 = 1$ point. So *A* loses 3 pairwise comparisons.

(b) Candidate *C*, with 3 points, is the winner. [The complete ranking is *C, D, B, E, A*.]

1.6. Fairness Criteria

51. First, we determine the winner using the Borda count.

A has $4 \times 6 + 3 \times 0 + 2 \times 0 + 1 \times (2 + 3) = 29$ points.

B has $4 \times 2 + 3 \times 6 + 2 \times 3 + 1 \times 0 = 32$ points.

C has $4 \times 3 + 3 \times 2 + 2 \times 6 + 1 \times 0 = 30$ points.

D has $4 \times 0 + 3 \times 3 + 2 \times 2 + 1 \times 6 = 19$ points.

So candidate *B* is the winner using Borda count. However, candidate *A* has 6 of the 11 votes (a majority) and beats all three other candidates (*B*, *C*, and *D*) in head-to-head comparisons. That is, candidate *A* is a Condorcet candidate. Since this candidate did not win using the Borda count, this is a violation of the Condorcet criterion.

53. The winner of this election is candidate *R* using the plurality method. Now *F* is clearly a nonwinning candidate. Removing *F* as a candidate leaves the following preference table.

Number of voters	49	48	3
1st choice	*R*	*H*	*H*
2nd choice	*H*	*S*	*S*
3rd choice	*O*	*O*	*O*
4th choice	*S*	*R*	*R*

In a recount, candidate *H* would be the winner using the plurality method. This is a violation of the IIA criterion.

55. First, we determine the winner using plurality-with-elimination.
Round 1:

Candidate	A	B	C	D	E
Number of first-place votes	8	3	5	5	0

Candidate *E* is eliminated.
Round 2: No votes are shifted.

Candidate	A	B	C	D	E
Number of first-place votes	8	3	5	5	

Candidate *B* is eliminated. Round 3: The 3 first-place votes for *B* now go to *A* (since *E* has been eliminated).

Candidate	A	B	C	D	E
Number of first-place votes	11		5	5	

Candidate *A* now has a majority (11 of the 21 votes) and is declared the winner. Now *C* is clearly a nonwinning (irrelevant) candidate. Removing *C* as a candidate leaves the following preference table.

Number of voters	5	5	3	3	3	2
1st choice	A	E	A	D	B	D
2nd choice	B	D	D	B	E	B
3rd choice	D	B	B	E	A	A
4th choice	E	A	E	A	D	E

We now recount using the plurality-with-elimination.

Round 1:

Candidate	A	B	D	E
Number of first-place votes	8	3	5	5

Candidate *B* is eliminated.
Round 2: The 3 first-place votes that went to *B* now shift to *E*.

Candidate	A	B	D	E
Number of first-place votes	8		5	8

Candidate *D* is eliminated.
Round 3: *D* had 5 first-place votes. Of these, 3 go to *E* and 2 go to *A* (since *B* was eliminated).

Candidate	A	B	D	E
Number of first-place votes	10			11

Candidate *E* now has a majority (11 of the 21 votes) and is declared the winner. Remember that candidate *A* was the winner before candidate *C* was removed. This is a violation of the IIA criterion.

57. If *X* is the Condorcet candidate, then by definition *X* wins every pairwise comparison and is, therefore, the winner under the method of pairwise comparisons.

59. When a voter moves a candidate up in his or her ballot, that candidate's Borda points increase. It follows that if *X* had the most Borda points and a voter changes his or her ballot to rank *X* higher, then *X* still has the most Borda points.

JOGGING

61. Suppose the two candidates are A and B and that A gets a first-place votes and B gets b first-place votes and suppose that $a > b$. Then A has a majority of the votes and the preference schedule is

Number of voters	a	b
1st choice	A	B
2nd choice	B	A

It is clear that candidate A wins the election under the plurality method, the plurality-with-elimination method, and the method of pairwise comparisons. Under the Borda count method, A gets $2a + b$ points while B gets $2b + a$ points. Since $a > b$, $2a + b > 2b + a$ and so again A wins the election.

63. The number of points under this variation is complementary to the number of points under the standard Borda count method: a first place is worth 1 point instead of N, a second place is worth 2 points instead of $N - 1, \ldots$, a last place is worth N points instead of 1. It follows that having the fewest points here is equivalent to having the most points under the standard Borda count method, having the second fewest is equivalent to having the second most, and so on.

As another way to see this, suppose candidates C_1 and C_2 receive p_1 and p_2 points respectively using the Borda count as originally described in the chapter and r_1 and r_2 points under the variation described in this exercise. Then, for an election with k voters, we have $p_1 + r_1 = p_2 + r_2 = k(N + 1)$. So if $p_1 < p_2$, we have $-p_1 > -p_2$ and so $k(N + 1) - p_1 > k(N + 1) - p_2$ which implies $r_1 > r_2$. Consequently the relative ranking of the candidates is not changed.

65. (a) Each of the voters gave Ohio State 25 points, so we compute 1625/25 to find 65 voters in the poll.

(b) Let $x =$ the number of Florida's second-place votes. Then $1529 = 24 \cdot x + 23 \cdot (65 - x)$ and so $x = 34$. This leaves $65 - 34 = 31$ third-place votes for Florida.

(c) Let $y =$ the number of Michigan's second-place votes. Then $1526 = 24 \cdot y + 23 \cdot (65 - y)$ and so $y = 31$. This leaves $65 - 31 = 34$ third-place votes for Michigan.

67. By looking at Dwayne Wade's vote totals, it is clear that 1 point is awarded for each third-place vote (0 points is too few and 2 points is too many for each third-place vote). Let $x =$ points awarded for each second-place vote. Then, $3x + 108 = 117$ gives $x = 3$ points for each second-place vote. Next, let $y =$ points awarded for each first-place vote. Looking at LeBron James' votes, we see that $78y + 39 \times 3 + 1 \times 1 = 508$. Solving this equation gives $y = 5$ points for each first-place vote.

69. (a) Round 1:

Number of voters	14	10	8	4	1
1st choice	A	C	D	B	C
2nd choice	B	B	C	D	D
3rd choice	C	D	B	C	B
4th choice	D	A	A	A	A

Candidate A with 23 last-place votes is eliminated.

Round 2:

Number of voters	14	10	8	4	1
1st choice	B	C	D	B	C
2nd choice	C	B	C	D	D
3rd choice	D	D	B	C	B

Candidate B has $8 + 1 = 9$ last-place votes, C has 4 last-place votes, and D has $14 + 10 = 24$ last-place votes. So candidate D is eliminated.

Round 3:

Number of voters	14	10	8	4	1
1st choice	B	C	C	B	C
2nd choice	C	B	B	C	B

B has more last-place votes (meaning C has more first-place votes) and C is declared the winner.

(b) Consider the election given by the preference schedule below. Here A is a Condorcet candidate but, having the most last-place votes, is eliminated in the first round.

Number of voters	10	6	6	3	3
1st choice	B	A	A	D	C
2nd choice	C	B	C	A	A
3rd choice	D	D	B	C	B
4th choice	A	C	D	B	D

(c) Consider the election given by the preference schedule below. Here B wins under the Coombs method (C is eliminated first and B is preferred to A by 17 voters). However, if 8 voters move B from their 3rd choice to their 2nd choice, then C wins (since A would be eliminated first and C is preferred head-to-head over B).

Number of voters	10	8	7	4
1st choice	B	C	C	A
2nd choice	A	A	B	B
3rd choice	C	B	A	C

RUNNING

71. One reasonable approach would be to use variables x_k, $1 \le k \le 10$, to represent the number of points handed out for kth place. With so many variables, there is some flexibility in determining the values of these variables. For starters, choose $x_{10} = 1$, $x_9 = 2$, $x_8 = 3$, $x_7 = 4$, $x_6 = 5$, and $x_5 = 6$. In that case, voting for Keuchel forces $3x_4 + 8 \times 6 + 5 \times 5 + 1 \times 4 + 3 \times 3 = 107$. That is, $x_4 = 7$. Then, voting for Machado forces $4x_3 + 11 \times 7 + 5 \times 6 + 1 \times 5 + 1 \times 4 + 1 \times 3 + 3 \times 2 + 1 \times 1 = 158$. So $x_3 = 8$. We check this these choices work for Cain. But $20x_3 + 8 \times 7 + 1 \times 5 + 1 \times 4 = 225$ also gives $x_3 = 8$. For Trout and Donaldson, we have two more equations that must hold:

$7x_1 + 22x_2 + 1 \times 8 = 304$

$23x_1 + 7x_2 = 385$

One technique would be to solve these two linear equations in two unknowns by solving one of the equations for one variable and substituting that expression into the other equation. Another method would be to "guess" that $x_2 = 9$ and note that $x_1 = 14$ solves both equations. The point values on each ballot are thus determined to be 14 points for 1st, 9 points for 2nd, 8 points for 3rd, 7 points for 4th, 6 points for 5th, 5 points for 6th, 4 points for 7th, 3 points for 8th, 2 points for 9th, and 1 point for 10th.

73. (a) The Math Club election serves as an example. Candidate *A* has a majority of last-place votes (23 of the 37) and yet wins the election when the plurality method is used.

Number of voters	14	10	8	4	1
1st choice	*A*	*C*	*D*	*B*	*C*
2nd choice	*B*	*B*	*C*	*D*	*D*
3rd choice	*C*	*D*	*B*	*C*	*B*
4th choice	*D*	*A*	*A*	*A*	*A*

(b) In this election, *C* is the winner under the plurality method (in the case of the tie in round 1, eliminate all candidates with the fewest number of votes). However, a majority of the voters (4 out of 7) prefer both *A* and *B* over *C*.

Number of voters	2	2	3
1st choice	*A*	*B*	*C*
2nd choice	*B*	*A*	*A*
3rd choice	*C*	*C*	*B*

(c) Under the method of pairwise comparisons, if a majority of voters have candidate *X* ranked last on their ballot, then candidate *X* will never win a head-to-head comparison (since any other candidate *Y* is preferred to *X* by a majority of voters). Thus, *X* will end up with no points under the method and cannot win the election.

(d) Suppose there are *N* candidates, *x* is the number of voters placing *X* last, and *y* is the remaining number of voters. Since a majority of the voters place *X* last, $x > y$. The *maximum* number of points that *X* can receive using a Borda count is $x + Ny$. (One point for each of the *x* last place votes and, assuming all other voters place *X* in first-place, *N* points for each of the other *y* voters.) The total number of points given out by each voter is $1 + 2 + 3 + ... + N = N(N + 1)/2$ and so the total number of points given out by all $x + y$ voters is $N(N + 1)(x + y)/2$. Since some candidate must receive at least 1/*N* of the total points, some candidate must receive at least $(N + 1)(x + y)/2$ points. Since $x > y$, we have $(N + 1)(x + y)/2 = (Nx + x + Ny + y)/2 > (Ny + x + Ny + y)/2 > x + Ny$, and consequently, *X* cannot be the winner of the election using the Borda count method.

75. Suppose there are *k* voters and *N* candidates.

Case 1: *k* is odd, say $k = 2t + 1$. Suppose the candidate with a majority of the first-place votes is *X*. The fewest possible Borda points *X* can have is $F(t + 1) + t$ [when there are $(t + 1)$ votes that place *X* first and the remaining votes place *X* last]. The most Borda points that any other candidate can have is $(N - 1)(t + 1) + Ft$ [when there are $(t + 1)$ voters that place the candidate second and the remaining voters place that candidate first]. Thus, the majority criterion will be satisfied when $F(t + 1) + t > (N - 1)(t + 1) + Ft$, which after simplification implies $F > N(t + 1) - (2t + 1)$, or $F > N\left(\frac{k+1}{2}\right) - k$.

Case 2: *k* is even, say $k = 2t$. An argument similar to the one given in case 1 gives the inequality $F(t + 1) + (t - 1) > (N - 1)(t + 1) + F(t - 1)$ which after simplification implies $F > [N(t + 1) - 2t]/2$, or $F > \left[N\left(\frac{k}{2}+1\right) - k\right]/2$.

Chapter 2

WALKING

2.1. Weighted Voting

1. (a) A generic weighted voting system with $N = 5$ players is described using notation $[q : w_1, w_2, w_3, w_4, w_5]$ where q represents the value of the quota and w_i represents the weight of player P_i. In this case, the players are the partners, $w_1 = 15$, $w_2 = 12$, $w_3 = w_4 = 10$, and $w_5 = 3$. Since the total number of votes is $15 + 12 + 10 + 10 + 3 = 50$ and the quota is determined by a simple majority (more than 50% of the total number of votes), $q = 26$. That is, the partnership can be described by $[26 : 15, 12, 10, 10, 3]$.

   (b) Since $\left(\frac{2}{3}\right)50 = 33\frac{1}{3}$, we choose the quota q as the smallest integer larger than this value which is 34. The partnership can thus be described by $[34 : 15, 12, 10, 10, 3]$.

3. (a) The quota must be *more than* half of the total number of votes. This system has $6 + 4 + 3 + 3 + 2 + 2 = 20$ total votes. Since 50% of 20 is 10, the smallest possible quota would be 11. Note: $q = 10$ is *not* sufficient. If 10 votes are cast in favor and 10 cast against a motion, that motion should not pass.

   (b) The largest value q can take is 20, the total number of votes.

   (c) $\frac{3}{4} \times 20 = 15$, so the value of the quota q would be 15.

   (d) The value of the quota q would be *strictly larger than* 15. That is, 16.

5. (a) P_1, the player with the most votes, does not have veto power since the other players combined have $3 + 3 + 2 = 8$ votes and can successfully pass a motion without him. The other players can't have veto power either then as they have fewer votes and hence less (or equal) power. So no players have veto power in this system.

   (b) P_1 does have veto power here since the other players combined have only $3 + 3 + 2 = 8$ votes and cannot successfully pass a motion without him. P_2, on the other hand, does not have veto power since the other players combined have the $4 + 3 + 2 = 9$ votes necessary to meet the quota. Since P_3 and P_4 have the same or fewer votes than P_2, it follows that only P_1 has veto power.

   (c) P_4, the player with the fewest votes, does not have veto power here since the other players combined have $4 + 3 + 3 = 10$ votes and can pass a motion without him. P_3, on the other hand, does have veto power since the other players combined have only $4 + 3 + 2 = 9$ votes. Since P_1 and P_2 have the same or more votes than P_3, it follows that P_1, P_2, and P_3 all have veto power.

   (d) P_4, the player with the fewest votes, has veto power since the other players combined have only $4 + 3 + 3 = 10$ votes and cannot pass a motion without him. Since the other players all have the same or more votes than P_4, it follows that all players have veto power in this system.

7. (a) In order for all three players to have veto power, the player having the fewest votes (the weakest) must have veto power. In order for that to happen, the quota q must be *strictly* larger than $7 + 5 = 12$. The smallest value of q for which this is true is $q = 13$.

(b) In order for P_2 to have veto power, the quota q must be strictly larger than $7 + 3 = 10$. The smallest value of q for which this is true is $q = 11$. [Note: When $q = 11$, we note that P_3 does not have veto power since the other two players have $7 + 5 = 12$ votes.]

9. To determine the number of votes each player has, write the weighted voting system as $[49: 4x, 2x, x, x]$.

(a) If the quota is defined as a simple majority of the votes, then x is the largest integer satisfying
$$49 > \frac{4x + 2x + x + x}{2}$$ which means that $8x < 98$ or $x < 12.25$. So, $x = 12$ and the system can be described as $[49: 48, 24, 12, 12]$.

(b) If the quota is defined as more than two-thirds of the votes, then x is the largest integer satisfying
$$49 > \frac{2(4x + 2x + x + x)}{3}$$ which means that $16x < 147$ or $x < 9.1875$. So, $x = 9$ and the system can be described as $[49: 36, 18, 9, 9]$.

(c) If the quota is defined as more than three-fourths of the votes, then x is the largest integer satisfying
$$49 > \frac{3(4x + 2x + x + x)}{4}$$ which means that $24x < 196$ or $x < 8.167$. So, $x = 8$ and the system can be described as $[49: 32, 16, 8, 8]$.

2.2. Banzhaf Power

11. (a) $w_1 + w_3 = 7 + 3 = 10$

(b) Since $(7 + 5 + 3)/2 = 7.5$, the smallest allowed value of the quota q in this system is 8. For $\{P_1, P_3\}$ to be a winning coalition, the quota q could be at most 10. So, the values of the quota for which $\{P_1, P_3\}$ is winning are 8, 9, and 10.

(c) Since $7 + 5 + 3 = 15$, the largest allowed value of the quota q in this system is 15. For $\{P_1, P_3\}$ to be a losing coalition, the quota q must be strictly greater than its weight, 10. So, the values of the quota for which $\{P_1, P_3\}$ is losing are 11, 12, 13, 14, and 15.

13. P_1 is critical (underlined) three times; P_2 is critical three times; P_3 is critical once ; P_4 is critical once. The total number of times the players are critical is 8 (number of underlines). The Banzhaf power distribution is $\beta_1 = 3/8$; $\beta_2 = 3/8$; $\beta_3 = 1/8$; $\beta_4 = 1/8$.

15. (a) P_1 is critical since the other players only have $5 + 2 = 7$ votes. P_2 is also critical since the other players only have $6 + 2 = 8$ votes. However, P_4 is not critical since the other two players have $6 + 5 = 11$ (more than $q = 10$).

(b) The winning coalitions are those whose weights are 10 or more. These are: $\{P_1, P_2\}$, $\{P_1, P_3\}$, $\{P_1, P_2, P_3\}$, $\{P_1, P_2, P_4\}$, $\{P_1, P_3, P_4\}$, $\{P_2, P_3, P_4\}$, $\{P_1, P_2, P_3, P_4\}$.

(c) We underline the critical players in each winning coalition: $\{\underline{P_1}, \underline{P_2}\}$, $\{\underline{P_1}, \underline{P_3}\}$, $\{\underline{P_1}, P_2, P_3\}$, $\{\underline{P_1}, \underline{P_2}, P_4\}$, $\{\underline{P_1}, \underline{P_3}, P_4\}$, $\{\underline{P_2}, \underline{P_3}, \underline{P_4}\}$, $\{P_1, P_2, P_3, P_4\}$.

Then, it follows that $\beta_1 = \dfrac{5}{12}$; $\beta_2 = \beta_3 = \dfrac{3}{12}$; $\beta_4 = \dfrac{1}{12}$.

17. (a) The winning coalitions (with critical players underlined) are: $\{\underline{P_1}, \underline{P_2}\}$, $\{\underline{P_1}, \underline{P_3}\}$, and $\{\underline{P_1}, P_2, P_3\}$. P_1 is critical three times; P_2 is critical one time; P_3 is critical one time. The total number of times the players are critical is 5. The Banzhaf power distribution is $\beta_1 = 3/5$; $\beta_2 = 1/5$; $\beta_3 = 1/5$.

(b) The winning coalitions (with critical players underlined) are: $\{\underline{P_1}, \underline{P_2}\}$, $\{\underline{P_1}, \underline{P_3}\}$, and $\{\underline{P_1}, P_2, P_3\}$. P_1 is critical three times; P_2 is critical one time; P_3 is critical one time. The total number of times the players are critical is 5. The Banzhaf power distribution is $\beta_1 = 3/5$; $\beta_2 = 1/5$; $\beta_3 = 1/5$. The distributions in (a) and (b) are the same.

19. (a) The winning coalitions (with critical players underlined) are: $\{\underline{P_1}, \underline{P_2}, \underline{P_3}\}$, $\{\underline{P_1}, \underline{P_2}, \underline{P_4}\}$, $\{\underline{P_1}, \underline{P_2}, \underline{P_5}\}$, $\{\underline{P_1}, \underline{P_3}, \underline{P_4}\}$, $\{\underline{P_1}, P_2, P_3, P_4\}$, $\{\underline{P_1}, P_2, P_3, P_5\}$, $\{\underline{P_1}, P_2, P_4, P_5\}$, $\{\underline{P_1}, P_3, P_4, P_5\}$, $\{\underline{P_2}, P_3, P_4, P_5\}$, $\{P_1, P_2, P_3, P_4, P_5\}$. The Banzhaf power distribution is $\beta_1 = 8/24$; $\beta_2 = 6/24$; $\beta_3 = 4/24$; $\beta_4 = 4/24$; $\beta_5 = 2/24$.

(b) The situation is like (a) except that $\{P_1, P_2, P_5\}$, $\{P_1, P_3, P_4\}$ and $\{P_2, P_3, P_4, P_5\}$ are now losing coalitions. In addition, P_2 is now critical in $\{P_1, P_2, P_3, P_4\}$, P_3 is now critical in $\{P_1, P_2, P_3, P_5\}$, P_4 is now critical in $\{P_1, P_2, P_4, P_5\}$, P_5 is now critical in $\{P_1, P_3, P_4, P_5\}$, and P_1 is now critical in the grand coalition $\{P_1, P_2, P_3, P_4, P_5\}$. The winning coalitions (with critical players underlined) are now: $\{\underline{P_1}, \underline{P_2}, \underline{P_3}\}$, $\{\underline{P_1}, \underline{P_2}, \underline{P_4}\}$, $\{\underline{P_1}, \underline{P_2}, P_3, P_4\}$, $\{\underline{P_1}, \underline{P_2}, \underline{P_3}, P_5\}$, $\{\underline{P_1}, \underline{P_2}, \underline{P_4}, P_5\}$, $\{\underline{P_1}, \underline{P_3}, \underline{P_4}, \underline{P_5}\}$, $\{\underline{P_1}, P_2, P_3, P_4, P_5\}$. The Banzhaf power distribution is $\beta_1 = 7/19$; $\beta_2 = 5/19$; $\beta_3 = 3/19$; $\beta_4 = 3/19$; $\beta_5 = 1/19$.

(c) This situation is like (b) with the following exceptions: $\{P_1, P_2, P_4\}$ and $\{P_1, P_3, P_4, P_5\}$ are now losing coalitions; P_3 is critical in $\{P_1, P_2, P_3, P_4\}$; P_5 is critical in $\{P_1, P_2, P_4, P_5\}$; and P_2 is critical in $\{P_1, P_2, P_3, P_4, P_5\}$. The winning coalitions (with critical players underlined) are now: $\{\underline{P_1}, \underline{P_2}, \underline{P_3}\}$, $\{\underline{P_1}, \underline{P_2}, \underline{P_3}, P_4\}$, $\{\underline{P_1}, \underline{P_2}, \underline{P_3}, P_5\}$, $\{\underline{P_1}, \underline{P_2}, \underline{P_4}, P_5\}$, $\{\underline{P_1}, \underline{P_2}, P_3, P_4, P_5\}$. The Banzhaf power distribution is $\beta_1 = 5/15$; $\beta_2 = 5/15$; $\beta_3 = 3/15$; $\beta_4 = 1/15$; $\beta_5 = 1/15$.

(d) Since the quota equals the total number of votes in the system, the only winning coalition is the grand coalition and every player is critical in that coalition. The Banzhaf power distribution is easy to calculate in the case since all players share power equally. It is $\beta_1 = 1/5$; $\beta_2 = 1/5$; $\beta_3 = 1/5$; $\beta_4 = 1/5$; $\beta_5 = 1/5$.

21. (a) A player is critical in a coalition if that coalition without the player is not on the list of winning coalitions. So, in this case, the critical players are underlined below.
$\{\underline{P_1}, \underline{P_2}\}$, $\{\underline{P_1}, \underline{P_3}\}$, $\{\underline{P_1}, P_2, P_3\}$

(b) P_1 is critical three times; P_2 is critical one time; P_3 is critical one time. The total number of times all players are critical is $3 + 1 + 1 = 5$. The Banzhaf power distribution is $\beta_1 = 3/5$; $\beta_2 = 1/5$; $\beta_3 = 1/5$.

23. (a) The winning coalitions (with critical players underlined) are $\{\underline{P_1}, \underline{P_2}\}$, $\{\underline{P_1}, \underline{P_3}\}$, $\{\underline{P_2}, \underline{P_3}\}$, $\{P_1, P_2, P_3\}$, $\{\underline{P_1}, \underline{P_2}, P_4\}$, $\{\underline{P_1}, \underline{P_2}, P_5\}$, $\{\underline{P_1}, \underline{P_2}, P_6\}$, $\{\underline{P_1}, \underline{P_3}, P_4\}$, $\{\underline{P_1}, \underline{P_3}, P_5\}$, $\{\underline{P_1}, \underline{P_3}, P_6\}$, $\{\underline{P_2}, \underline{P_3}, P_4\}$, $\{\underline{P_2}, \underline{P_3}, P_5\}$, $\{\underline{P_2}, \underline{P_3}, P_6\}$.

(b) These winning coalitions (with critical players underlined) are $\{\underline{P_1}, \underline{P_2}, P_4\}$, $\{\underline{P_1}, \underline{P_3}, P_4\}$, $\{\underline{P_2}, \underline{P_3}, P_4\}$, $\{P_1, P_2, P_3, P_4\}$, $\{\underline{P_1}, \underline{P_2}, P_4, P_5\}$, $\{\underline{P_1}, \underline{P_2}, P_4, P_6\}$, $\{\underline{P_1}, \underline{P_3}, P_4, P_5\}$, $\{\underline{P_1}, \underline{P_3}, P_4, P_6\}$, $\{\underline{P_2}, \underline{P_3}, P_4, P_5\}$,

$\{\underline{P_2}, \underline{P_3}, P_4, P_6\}$, $\{P_1, P_2, P_3, P_4, P_5\}$, $\{P_1, P_2, P_3, P_4, P_6\}$, $\{\underline{P_1}, \underline{P_2}, P_4, P_5, P_6\}$,
$\{\underline{P_1}, \underline{P_3}, P_4, P_5, P_6\}$, $\{\underline{P_2}, \underline{P_3}, P_4, P_5, P_6\}$, $\{P_1, P_2, P_3, P_4, P_5, P_6\}$.

(c) P_4 is never a critical player since every time it is part of a winning coalition, that coalition is a winning coalition without P_4 as well. So, $\beta_4 = 0$.

(d) A similar argument to that used in part (c) shows that P_5 and P_6 are also dummies. One could also argue that any player with fewer votes than P_4, a dummy, will also be a dummy. So, P_4, P_5, and P_6 will never be critical -- they all have zero power.

The only winning coalitions with only two players are $\{P_1, P_2\}$, $\{P_1, P_3\}$, and $\{P_2, P_3\}$; and both players are critical in each of those coalitions. All other winning coalitions consist of one of these coalitions plus additional players, and the only critical players will be the ones from the two-player coalition. So P_1, P_2, and P_3 will be critical in every winning coalition they are in, and they will all be in the same number of winning coalitions, so they all have the same power. Thus, the Banzhaf power distribution is $\beta_1 = 1/3$; $\beta_2 = 1/3$; $\beta_3 = 1/3$; $\beta_4 = 0$; $\beta_5 = 0$; $\beta_6 = 0$.

25. (a) $\{\underline{A}, \underline{B}\}, \{\underline{A}, \underline{C}\}, \{\underline{B}, \underline{C}\}, \{A, B, C\}, \{\underline{A}, \underline{B}, D\}, \{\underline{A}, \underline{C}, D\}, \{\underline{B}, \underline{C}, D\}, \{A, B, C, D\}$

(b) A, B, and C have Banzhaf power index of 4/12 each; D is a dummy. (D is never a critical player, and the other three clearly have equal power.)

2.3. Shapley-Shubik Power

27. P_1 is pivotal (underlined) ten times; P_2 is pivotal (again, underlined) ten times; P_3 is pivotal twice ; P_4 is pivotal twice. The total number of times the players are pivotal is $4! = 24$ (number of underlines). The Shapley-Shubik power distribution is $\sigma_1 = 10/24$; $\sigma_2 = 10/24$; $\sigma_3 = 2/24$; $\sigma_4 = 2/24$.

29. (a) There are $3! = 6$ sequential coalitions of the three players. Each pivotal player is underlined.
$< P_1, \underline{P_2}, P_3 >, < P_1, \underline{P_3}, P_2 >, < P_2, \underline{P_1}, P_3 >, < P_2, P_3, \underline{P_1} >, < P_3, \underline{P_1}, P_2 >, < P_3, P_2, \underline{P_1} >$

(b) P_1 is pivotal four times; P_2 is pivotal one time; P_3 is pivotal one time. The Shapley-Shubik power distribution is $\sigma_1 = 4/6$; $\sigma_2 = 1/6$; $\sigma_3 = 1/6$.

31. (a) Since P_1 is a dictator, $\sigma_1 = 1$, $\sigma_2 = 0$, $\sigma_3 = 0$, and $\sigma_4 = 0$.

(b) There are $4! = 24$ sequential coalitions of the four players. Each pivotal player is underlined.
$< P_1, \underline{P_2}, P_3, P_4 >, < P_1, \underline{P_2}, P_4, P_3 >, < P_1, \underline{P_3}, P_2, P_4 >, < P_1, \underline{P_3}, P_4, P_2 >,$
$< P_1, P_4, \underline{P_2}, P_3 >, < P_1, P_4, \underline{P_3}, P_2 >, < P_2, \underline{P_1}, P_3, P_4 >, < P_2, \underline{P_1}, P_4, P_3 >,$
$< P_2, P_3, \underline{P_1}, P_4 >, < P_2, P_3, P_4, \underline{P_1} >, < P_2, P_4, \underline{P_1}, P_3 >, < P_2, P_4, P_3, \underline{P_1} >,$
$< P_3, \underline{P_1}, P_2, P_4 >, < P_3, \underline{P_1}, P_4, P_2 >, < P_3, P_2, \underline{P_1}, P_4 >, < P_3, P_2, P_4, \underline{P_1} >,$
$< P_3, P_4, \underline{P_1}, P_2 >, < P_3, P_4, P_2, \underline{P_1} >, < P_4, P_1, \underline{P_2}, P_3 >, < P_4, P_1, \underline{P_3}, P_2 >,$
$< P_4, P_2, \underline{P_1}, P_3 >, < P_4, P_2, P_3, \underline{P_1} >, < P_4, P_3, \underline{P_1}, P_2 >, < P_4, P_3, P_2, \underline{P_1} >$

P_1 is pivotal 16 times; P_2 is pivotal 4 times; P_3 is pivotal 4 times; P_4 is pivotal 0 times. The Shapley-Shubik power distribution is $\sigma_1 = 16/24$; $\sigma_2 = 4/24$; $\sigma_3 = 4/24$; $\sigma_4 = 0$.

(c) The only way a motion will pass is if P_1 and P_2 both support it. In fact, the second of these players that appears in a sequential coalition will be the pivotal player in that coalition. It follows that $\sigma_1 = 12/24 = 1/2$, $\sigma_2 = 12/24 = 1/2$, $\sigma_3 = 0$, and $\sigma_4 = 0$.

(d) Because the quota is so high, the only way a motion will pass is if P_1, P_2, and P_3 all support it. In fact, the third of these players that appears in a sequential coalition will always be the pivotal player in that coalition. It follows that $\sigma_1 = 8/24 = 1/3$, $\sigma_2 = 8/24 = 1/3$, $\sigma_3 = 8/24 = 1/3$, and $\sigma_4 = 0$.

33. (a) There are $4! = 24$ sequential coalitions of the four players. Each pivotal player is underlined.

$< P_1, \underline{P_2}, P_3, P_4 >$, $< P_1, \underline{P_2}, P_4, P_3 >$, $< P_1, \underline{P_3}, P_2, P_4 >$, $< P_1, \underline{P_3}, P_4, P_2 >$,

$< P_1, P_4, \underline{P_2}, P_3 >$, $< P_1, P_4, \underline{P_3}, P_2 >$, $< P_2, \underline{P_1}, P_3, P_4 >$, $< P_2, \underline{P_1}, P_4, P_3 >$,

$< P_2, P_3, \underline{P_1}, P_4 >$, $< P_2, P_3, \underline{P_4}, P_1 >$, $< P_2, P_4, \underline{P_1}, P_3 >$, $< P_2, P_4, \underline{P_3}, P_1 >$,

$< P_3, \underline{P_1}, P_2, P_4 >$, $< P_3, \underline{P_1}, P_4, P_2 >$, $< P_3, P_2, \underline{P_1}, P_4 >$, $< P_3, P_2, \underline{P_4}, P_1 >$,

$< P_3, P_4, \underline{P_1}, P_2 >$, $< P_3, P_4, \underline{P_2}, P_1 >$, $< P_4, P_1, \underline{P_2}, P_3 >$, $< P_4, P_1, \underline{P_3}, P_2 >$,

$< P_4, P_2, \underline{P_1}, P_3 >$, $< P_4, P_2, \underline{P_3}, P_1 >$, $< P_4, P_3, \underline{P_1}, P_2 >$, $< P_4, P_3, \underline{P_2}, P_1 >$

P_1 is pivotal in 10 coalitions; P_2 is pivotal in six coalitions; P_3 is pivotal in six coalitions; P_4 is pivotal in two coalitions. The Shapley-Shubik power distribution is $\sigma_1 = 10/24$; $\sigma_2 = 6/24$; $\sigma_3 = 6/24$; $\sigma_4 = 2/24$.

(b) This is the same situation as in (a) – there is essentially no difference between 51 and 59 because the players' votes are all multiples of 10. The Shapley-Shubik power distribution is thus still $\sigma_1 = 10/24$; $\sigma_2 = 6/24$; $\sigma_3 = 6/24$; $\sigma_4 = 2/24$.

(c) This is also the same situation as in (a) – any time a group of players has 51 votes, they must have 60 votes. The Shapley-Shubik power distribution is thus still $\sigma_1 = 10/24$; $\sigma_2 = 6/24$; $\sigma_3 = 6/24$; $\sigma_4 = 2/24$.

35. (a) There are $3! = 6$ sequential coalitions of the three players.

$< P_1, \underline{P_2}, P_3 >$, $< P_1, \underline{P_3}, P_2 >$, $< P_2, \underline{P_1}, P_3 >$, $< P_2, P_3, \underline{P_1} >$, $< P_3, \underline{P_1}, P_2 >$, $< P_3, P_2, \underline{P_1} >$

(The second player listed will always be pivotal unless the first two players listed are P_2 and P_3 since that is not a winning coalition. In that case, P_1 is pivotal.)

(b) P_1 is pivotal four times; P_2 is pivotal one time; P_3 is pivotal one time. The Shapley-Shubik power distribution is $\sigma_1 = 4/6$; $\sigma_2 = 1/6$; $\sigma_3 = 1/6$.

37. We proceed to identify pivotal players by moving down each column. We start with the first column. Since P_2 is pivotal in the sequential coalition $< P_1, \underline{P_2}, P_3, P_4 >$, we know that $\{P_1, P_2\}$ form a winning coalition. This tells us that P_2 is also pivotal in the second sequential coalition $< P_1, \underline{P_2}, P_4, P_3 >$ listed in the first column. A similar argument tells us that P_3 is pivotal in the fourth sequential coalition $< P_1, \underline{P_3}, P_4, P_2 >$ listed in the first column.

In the second column, P_1 is clearly pivotal in the sequential coalition $< P_2, \underline{P_1}, P_3, P_4 >$ since we know that $\{P_1, P_2\}$ form a winning coalition. Similarly, P_1 is pivotal in the sequential coalition $< P_2, \underline{P_1}, P_4, P_3 >$.

In the third column, P_1 is pivotal in the sequential coalition $< P_3, \underline{P_1}, P_2, P_4 >$ since the third sequential coalition in the first column ($< P_1, \underline{P_3}, P_2, P_4 >$) identified $\{P_1, P_3\}$ as a winning coalition. Similarly, P_1 is pivotal in the sequential coalition $< P_3, \underline{P_1}, P_4, P_2 >$. Since P_1 was pivotal in the sequential coalition $< P_2, P_3, \underline{P_1}, P_4 >$ back in the second column, it must also be the case that P_1 is pivotal in the sequential coalition $< P_3, P_2, \underline{P_1}, P_4 >$. Similarly, P_4 is pivotal in the sequential coalition $< P_3, P_2, \underline{P_4}, P_1 >$ since P_4 was also pivotal in the sequential coalition $< P_2, P_3, \underline{P_4}, P_1 >$ back in the second column.

In the fourth column, the third player listed will always be pivotal since the first two players when listed first in the opposite order (in an earlier column) never contain a pivotal player and the fourth player was never pivotal in the first three columns. The final listing of pivotal players is found below.

$< P_1, \underline{P_2}, P_3, P_4 >, < P_2, \underline{P_1}, P_3, P_4 >, < P_3, \underline{P_1}, P_2, P_4 >, < P_4, P_1, \underline{P_2}, P_3 >,$

$< P_1, \underline{P_2}, P_4, P_3 >, < P_2, \underline{P_1}, P_4, P_3 >, < P_3, \underline{P_1}, P_4, P_2 >, < P_4, P_1, \underline{P_3}, P_2 >,$

$< P_1, \underline{P_3}, P_2, P_4 >, < P_2, P_3, \underline{P_1}, P_4 >, < P_3, P_2, \underline{P_1}, P_4 >, < P_4, P_2, \underline{P_1}, P_3 >,$

$< P_1, \underline{P_3}, P_4, P_2 >, < P_2, P_3, \underline{P_4}, P_1 >, < P_3, P_2, \underline{P_4}, P_1 >, < P_4, P_2, \underline{P_3}, P_1 >,$

$< P_1, P_4, \underline{P_2}, P_3 >, < P_2, P_4, \underline{P_1}, P_3 >, < P_3, P_4, \underline{P_1}, P_2 >, < P_4, P_3, \underline{P_1}, P_2 >,$

$< P_1, P_4, \underline{P_3}, P_2 >, < P_2, P_4, \underline{P_3}, P_1 >, < P_3, P_4, \underline{P_2}, P_1 >, < P_4, P_3, \underline{P_2}, P_1 >$

The Shapley-Shubik distribution is then $\sigma_1 = 10/24$; $\sigma_2 = 6/24$; $\sigma_3 = 6/24$; $\sigma_4 = 2/24$.

2.4. Subsets and Permutations

39. (a) A set with N elements has 2^N subsets. So, A has $2^{10} = 1024$ subsets.

(b) There is only one subset, namely the empty set $\{\ \}$, having less than one element. So, using part (a), the number of subsets of A having one or more elements is $1024 - 1 = 1023$.

(c) Each element of A can be used to form a subset containing exactly one element. So, there are exactly 10 such subsets.

(d) The number of subsets of A having two or more elements is the number of subsets of A not having either 0 or 1 element. From parts (a), (b) and (c), we calculate this number as $1024 - 1 - 10 = 1013$.

41. (a) A weighted voting system with N players has $2^N - 1$ coalitions. So, a system with 10 players has $2^{10} - 1 = 1023$ coalitions.

(b) The number of coalitions with two or more players is all of the coalitions minus the number of those having at most one player. But we know that there are exactly 10 coalitions consisting of exactly one player. So, the number of coalitions with two or more players is $1023 - 10 = 1013$. See also Exercise 39(d).

43. (a) $2^6 - 1 = 63$ coalitions

(b) There are $2^5 - 1 = 31$ coalitions of the remaining five players P_2, P_3, P_4, P_5, and P_6. These are exactly those coalitions that do not include P_1.

(c) As in (b), there are $2^5 - 1 = 31$ coalitions of the remaining five players P_1, P_2, P_4, P_5, and P_6. These are exactly those coalitions that do not include P_3.

(d) There are $2^4 - 1 = 15$ coalitions of the remaining four players P_2, P_4, P_5, P_6. These are exactly those coalitions that do not include either P_1 or P_3.

(e) 16 coalitions include P_1 and P_3 (all $2^4 - 1 = 15$ coalitions of the remaining four players P_2, P_4, P_5, P_6 together with the empty coalition could be combined with these two players to form such a coalition).

45. **(a)** $13! = 6,227,020,800$

(b) $18! = 6,402,373,705,728,000 \approx 6.402374 \times 10^{15}$

(c) $25! = 15,511,210,043,330,985,984,000,000 \approx 1.551121 \times 10^{25}$

(d) There are $25!$ sequential coalitions of 25 players.

$$25! \text{ sequential coaltions} \times \frac{1 \text{ second}}{1,000,000,000,000 \text{ sequential coalitions}} \times \frac{1 \text{ hour}}{3600 \text{ seconds}} \times \frac{1 \text{ day}}{24 \text{ hours}} \times \frac{1 \text{ year}}{365 \text{ days}}$$

$\approx 491,857$ years

47. **(a)** $\dfrac{13!}{3!} = \dfrac{6,227,020,800}{6} = 1,037,836,800$

(b) $\dfrac{13!}{3!10!} = \dfrac{6,227,020,800}{6 \times 3,628,800} = 286$

(c) $\dfrac{13!}{4!9!} = \dfrac{6,227,020,800}{24 \times 362,880} = 715$

(d) $\dfrac{13!}{5!8!} = \dfrac{6,227,020,800}{120 \times 40,320} = 1287$

49. **(a)** $10! = 10 \times 9 \times 8 \times \ldots \times 3 \times 2 \times 1$
$= 10 \times 9!$

So, $9! = \dfrac{10!}{10} = \dfrac{3,628,800}{10} = 362,880$.

(b) $11! = 11 \times 10 \times 9 \times \ldots \times 3 \times 2 \times 1 = 11 \times 10!$

So, $\dfrac{11!}{10!} = \dfrac{11 \times 10!}{10!} = 11$.

(c) $11! = 11 \times 10 \times (9 \times \ldots \times 3 \times 2 \times 1) = 11 \times 10 \times 9!$

So, $\dfrac{11!}{9!} = \dfrac{11 \times 10 \times 9!}{9!} = 11 \times 10 = 110$.

(d) $\dfrac{9!}{6!} = \dfrac{9 \times 8 \times 7 \times 6!}{6!} = 9 \times 8 \times 7 = 504$

(e) $\dfrac{101!}{99!} = \dfrac{101 \times 100 \times 99!}{99!} = 101 \times 100 = 10,100$

51. (a) $7! = 5040$ sequential coalitions

(b) There are $6! = 720$ sequential coalitions of the remaining six players P_1, P_2, P_3, P_4, P_5 and P_6. These correspond exactly to those seven-player coalitions that have P_7 as the first player.

(c) As in (b), there are $6! = 720$ sequential coalitions of the remaining six players P_1, P_2, P_3, P_4, P_5 and P_6. These correspond exactly to those seven-player coalitions that have P_7 as the last player.

(d) $5040 - 720 = 4320$ sequential coalitions do not have P_1 listed as the first player.

53. (a) First, note that P_1 is pivotal in all sequential coalitions except when it is the first player. By 51(d), P_1 is pivotal in 4320 of the 5040 sequential coalitions.

(b) Based on the results of (a), $\sigma_1 = 4320/5040 = 6/7$.

(c) Since players P_2, P_3, P_4, P_5, P_6 and P_7 share the remaining power equally, $\sigma_1 = 4320/5040 = 6/7$,
$\sigma_2 = \sigma_3 = \sigma_4 = \sigma_5 = \sigma_6 = \sigma_7 = 120/5040 = 1/42$. [Note also that 1/42 is 1/7 divided into 6 equal parts.]

JOGGING

55. (a) Suppose that a winning coalition that contains P is not a winning coalition without P. Then P would be a critical player in that coalition, contradicting the fact that P is a dummy.

(b) P is a dummy $\Leftrightarrow$ P is never critical $\Leftrightarrow$ the numerator of its Banzhaf power index is 0 $\Leftrightarrow$ it's Banzhaf power index is 0.

(c) Suppose P is not a dummy. Then, P is critical in some winning coalition. Let S denote the other players in that winning coalition. The sequential coalition with the players in S first (in any order), followed by P and then followed by the remaining players has P as its pivotal player. Thus, P's Shapley-Shubik power index is not 0. Conversely, if P's Shapley-Shubik power index is not 0, then P is pivotal in some sequential coalition. A coalition consisting of P together with the players preceding P in that sequential coalition is a winning coalition and P is a critical player in it. Thus, P is not a dummy.

57. (a) The quota must be at least half of the total number of votes and not more than the total number of votes. $7 \le q \le 13$.

(b) For $q = 7$ or $q = 8$, P_1 is a dictator because $\{P_1\}$ is a winning coalition.

(c) For $q = 9$, only P_1 has veto power since P_2 and P_3 together have just 5 votes.

(d) For $10 \le q \le 12$, both P_1 and P_2 have veto power since no motion can pass without both of their votes. For $q = 13$, all three players have veto power.

(e) For $q = 7$ or $q = 8$, both P_2 and P_3 are dummies because P_1 is a dictator. For $10 \le q \le 12$, P_3 is a dummy since all winning coalitions contain $\{P_1, P_2\}$ which is itself a winning coalition.

59. (a) The winning coalitions for both weighted voting systems [8: 5, 3, 2] and [2: 1, 1, 0] are $\{P_1, P_2\}$ and $\{P_1, P_2, P_3\}$.

(b) The winning coalitions for both weighted voting systems [7: 4, 3, 2, 1] and [5: 3, 2, 1, 1] are $\{P_1, P_2\}$, $\{P_1, P_2, P_3\}$, $\{P_1, P_2, P_4\}$, $\{P_1, P_3, P_4\}$, and $\{P_1, P_2, P_3, P_4\}$.

(c) The winning coalitions for both weighted voting systems are those consisting of any three of the five players, any four of the five players, and the grand coalition.

(d) If a player is critical in a winning coalition, then that coalition is no longer winning if the player is removed. An equivalent system (with the same winning coalitions) will find that player critical in the same coalitions. When calculating the Banzhaf power indexes in two equivalent systems, the same players will be critical in the same coalitions. Thus, the numerators and denominators in each player's Banzhaf power index will be the same.

(e) If a player, P, is pivotal in a sequential coalition, then the players to P's left in the sequential coalition do not form a winning coalition, but including P does make it a winning coalition. An equivalent system (with the same winning coalitions) will find the player P pivotal in that same sequential coalition. When calculating the Shapley-Shubik indexes in two equivalent systems, the same players will be pivotal in the same sequential coalitions. Thus, the numerators and denominators in each Shapley-Shubik index will be the same.

61. (a) $V/2 < q \leq V - w_1$ (where V denotes the sum of all the weights). The idea here is that if P_1 does not have veto power, neither will any of the other players (since weights decrease in value).

(b) $V - w_N < q \leq V$; The idea here is that if P_N has veto power, so will all the other players.

(c) $V - w_i < q \leq V - w_{i+1}$

63. You should buy your vote from P_3. The following table explains why.

Buying a vote from	Resulting weighted voting system	Resulting Banzhaf power distribution	Your power
P_1	[8: 5, 4, 2, 2]	$\beta_1 = \frac{4}{12}$; $\beta_2 = \frac{4}{12}$; $\beta_3 = \frac{2}{12}$; $\beta_4 = \frac{2}{12}$	$\frac{2}{12}$
P_2	[8: 6, 3, 2, 2]	$\beta_1 = \frac{7}{10}$; $\beta_2 = \frac{1}{10}$; $\beta_3 = \frac{1}{10}$; $\beta_4 = \frac{1}{10}$	$\frac{1}{10}$
P_3	[8: 6, 4, 1, 2]	$\beta_1 = \frac{6}{10}$; $\beta_2 = \frac{2}{10}$; $\beta_3 = 0$; $\beta_4 = \frac{2}{10}$	$\frac{2}{10}$

65. (a) You should buy your vote from P_2. The following table explains why.

Buying a vote from	Resulting weighted voting system	Resulting Banzhaf power distribution	Your power
P_1	[18: 9, 8, 6, 4, 3]	$\beta_1 = \frac{4}{13}$; $\beta_2 = \frac{3}{13}$; $\beta_3 = \frac{3}{13}$; $\beta_4 = \frac{2}{13}$; $\beta_5 = \frac{1}{13}$	$\frac{1}{13}$
P_2	[18: 10, 7, 6, 4, 3]	$\beta_1 = \frac{9}{25}$; $\beta_2 = \frac{1}{5}$; $\beta_3 = \frac{1}{5}$; $\beta_4 = \frac{3}{25}$; $\beta_5 = \frac{3}{25}$	$\frac{3}{25}$
P_3	[18: 10, 8, 5, 4, 3]	$\beta_1 = \frac{5}{12}$; $\beta_2 = \frac{1}{4}$; $\beta_3 = \frac{1}{6}$; $\beta_4 = \frac{1}{12}$; $\beta_5 = \frac{1}{12}$	$\frac{1}{12}$
P_4	[18: 10, 8, 6, 3, 3]	$\beta_1 = \frac{5}{12}$; $\beta_2 = \frac{1}{4}$; $\beta_3 = \frac{1}{6}$; $\beta_4 = \frac{1}{12}$; $\beta_5 = \frac{1}{12}$	$\frac{1}{12}$

(b) You should buy 2 votes from P_2. The following table explains why.

Buying a vote from	Resulting weighted voting system	Resulting Banzhaf power distribution	Your power
P_1	[18: 8, 8, 6, 4, 4]	$\beta_1 = \frac{7}{27}$; $\beta_2 = \frac{7}{27}$; $\beta_3 = \frac{7}{27}$; $\beta_4 = \frac{1}{9}$; $\beta_5 = \frac{1}{9}$	$\frac{1}{9}$
P_2	[18: 10, 6, 6, 4, 4]	$\beta_1 = \frac{5}{13}$; $\beta_2 = \frac{2}{13}$; $\beta_3 = \frac{2}{13}$; $\beta_4 = \frac{2}{13}$; $\beta_5 = \frac{2}{13}$	$\frac{2}{13}$
P_3	[18: 10, 8, 4, 4, 4]	$\beta_1 = \frac{11}{25}$; $\beta_2 = \frac{1}{5}$; $\beta_3 = \frac{3}{25}$; $\beta_4 = \frac{3}{25}$; $\beta_5 = \frac{3}{25}$	$\frac{3}{25}$
P_4	[18: 10, 8, 6, 2, 4]	$\beta_1 = \frac{9}{25}$; $\beta_2 = \frac{7}{25}$; $\beta_3 = \frac{1}{5}$; $\beta_4 = \frac{1}{25}$; $\beta_5 = \frac{3}{25}$	$\frac{3}{25}$

(c) Buying a single vote from P_2 raises your power from $\frac{1}{25} = 4\%$ to $\frac{3}{25} = 12\%$. Buying a second vote

from P_2 raises your power to $\frac{2}{13} \approx 15.4\%$. The increase in power is less with the second vote, but if you

value power over money, it might still be worth it to you to buy that second vote.

67. (a) [24: 14, 8, 6, 4] is just [12: 7, 4, 3, 2] with each value multiplied by 2. Both have Banzhaf power distribution $\beta_1 = 2/5$; $\beta_2 = 1/5$; $\beta_3 = 1/5$; $\beta_4 = 1/5$.

(b) In the weighted voting system $[q : w_1, w_2, ..., w_N]$, if P_k is critical in a coalition then the sum of the weights of all the players in that coalition (including P_k) is at least q, but the sum of the weights of all the players in the coalition except P_k is less than q. Consequently, if the weights of all the players in that coalition are multiplied by $c > 0$ ($c = 0$ would make no sense), then the sum of the weights of all the players in the coalition (including P_k) is at least cq but the sum of the weights of all the players in the coalition except P_k is less than cq. Therefore P_k is critical in the same coalition in the weighted voting system $[cq : cw_1, cw_2, ..., cw_N]$. Since the critical players are the same in both weighted voting systems, the Banzhaf power distributions will be the same.

69. (a) A player is critical in a coalition if that coalition without the player is not on the list of winning coalitions. The critical players in the winning coalitions are underlined: $\{\underline{P_1}, \underline{P_2}, \underline{P_3}\}$, $\{\underline{P_1}, \underline{P_2}, \underline{P_4}\}$, $\{\underline{P_1}, \underline{P_3}, \underline{P_4}\}$, $\{\underline{P_1}, P_2, P_3, P_4\}$. So the Banzhaf power distribution is $\beta_1 = 4/10$, $\beta_2 = \beta_3 = \beta_4 = 2/10$.

(b) A player P is pivotal in a sequential coalition if the players that appear before P in that sequential coalition do not, as a group, appear on the list of winning coalitions, but would appear on the list of winning coalitions if P were included. For example, P_3 is pivotal in $< P_1, P_2, P_3, P_4 >$ since $\{P_1, P_2\}$ is not on the list of winning coalitions but $\{P_1, P_2, P_3\}$ is on that list. The pivotal players in each sequential coalition is underlined:
$< P_1, P_2, \underline{P_3}, P_4 >$, $< P_1, P_2, \underline{P_4}, P_3 >$, $< P_1, P_3, \underline{P_2}, P_4 >$, $< P_1, P_3, \underline{P_4}, P_2 >$,
$< P_1, P_4, \underline{P_2}, P_3 >$, $< P_1, P_4, \underline{P_3}, P_2 >$, $< P_2, P_1, \underline{P_3}, P_4 >$, $< P_2, P_1, \underline{P_4}, P_3 >$,
$< P_2, P_3, \underline{P_1}, P_4 >$, $< P_2, P_3, P_4, \underline{P_1} >$, $< P_2, P_4, \underline{P_1}, P_3 >$, $< P_2, P_4, P_3, \underline{P_1} >$,
$< P_3, P_1, \underline{P_2}, P_4 >$, $< P_3, P_1, \underline{P_4}, P_2 >$, $< P_3, P_2, \underline{P_1}, P_4 >$, $< P_3, P_2, P_4, \underline{P_1} >$,
$< P_3, P_4, \underline{P_1}, P_2 >$, $< P_3, P_4, P_2, \underline{P_1} >$, $< P_4, P_1, \underline{P_2}, P_3 >$, $< P_4, P_1, \underline{P_3}, P_2 >$,
$< P_4, P_2, \underline{P_1}, P_3 >$, $< P_4, P_2, P_3, \underline{P_1} >$, $< P_4, P_3, \underline{P_1}, P_2 >$, $< P_4, P_3, P_2, \underline{P_1} >$
So, the Shapley-Shubik power indices are $\sigma_1 = 12/24 = 1/2$, $\sigma_2 = \sigma_3 = \sigma_4 = 4/24 = 1/6$.

RUNNING

71. Write the weighted voting system as $[q: 8x, 4x, 2x, x]$. The total number of votes is $15x$. If x is even, then so

is $15x$. Since the quota is a simple majority, $q = \dfrac{15x}{2} + 1$. But, $\dfrac{15x}{2} + 1 = 7.5x + 1 \leq 8x$ since $x \geq 2$. So, P_1

(having $8x$ votes) is a dictator. If x is odd, then so is $15x$. Since the quota is a simple majority, $q = \dfrac{15x+1}{2}$.

But, $\dfrac{15x+1}{2} = 7.5x + \dfrac{1}{2} \leq 8x$ since $x \geq 1$. So, again, P_1 (having $8x$ votes) is a dictator.

73. (a) $[4: 2, 1, 1, 1]$ or $[9: 5, 2, 2, 2]$ are among the possible answers.

(b) The sequential coalitions (with pivotal players underlined) are:

$< H, A_1, \underline{A_2}, A_3 >, < H, A_1, \underline{A_3}, A_2 >, < H, A_2, \underline{A_1}, A_3 >, < H, A_2, \underline{A_3}, A_1 >,$

$< H, A_3, \underline{A_1}, A_2 >, < H, A_3, \underline{A_2}, A_1 >, < A_1, H, \underline{A_2}, A_3 >, < A_1, H, \underline{A_3}, A_2 >,$

$< A_1, A_2, \underline{H}, A_3 >, < A_1, A_2, A_3, \underline{H} >, < A_1, A_3, \underline{H}, A_2 >, < A_1, A_3, A_2, \underline{H} >,$

$< A_2, H, \underline{A_1}, A_3 >, < A_2, H, \underline{A_3}, A_1 >, < A_2, A_1, \underline{H}, A_3 >, < A_2, A_1, A_3, \underline{H} >,$

$< A_2, A_3, \underline{H}, A_1 >, < A_2, A_3, A_1, \underline{H} >, < A_3, H, \underline{A_1}, A_2 >, < A_3, H, \underline{A_2}, A_1 >,$

$< A_3, A_1, \underline{H}, A_2 >, < A_3, A_1, A_2, \underline{H} >, < A_3, A_2, \underline{H}, A_1 >, < A_3, A_2, A_1, \underline{H} >.$

H is pivotal in 12 coalitions; A_1 is pivotal in four coalitions; A_2 is pivotal in four coalitions; A_3 is pivotal in four coalitions. The Shapley-Shubik power distribution is $\sigma_H = 12/24 = 1/2$;

$\sigma_{A_1} = \sigma_{A_2} = \sigma_{A_3} = 4/24 = 1/6$.

75. (a) The losing coalitions are $\{P_1\}$, $\{P_2\}$, and $\{P_3\}$. The complements of these coalitions are $\{P_2, P_3\}$, $\{P_1, P_3\}$, and $\{P_1, P_2\}$ respectively, all of which are winning coalitions.

(b) The losing coalitions are $\{P_1\}$, $\{P_2\}$, $\{P_3\}$, $\{P_4\}$, $\{P_2, P_3\}$, $\{P_2, P_4\}$, and $\{P_3, P_4\}$. The complements of these coalitions are $\{P_2, P_3, P_4\}$, $\{P_1, P_3, P_4\}$, $\{P_1, P_2, P_4\}$, $\{P_1, P_2, P_3\}$, $\{P_1, P_4\}$, $\{P_1, P_3\}$, and $\{P_1, P_2\}$ respectively, all of which are winning coalitions.

(c) If P is a dictator, the losing coalitions are all the coalitions without P; the winning coalitions are all the coalitions that include P. The complement of any coalition without P (losing) is a coalition with P (winning).

(d) Take the grand coalition out of the picture for a moment. Of the remaining $2^N - 2$ coalitions, half are losing coalitions and half are winning coalitions, since each losing coalition pairs up with a winning coalition (its complement). Half of $2^N - 2$ is $2^{N-1} - 1$. In addition, we have the grand coalition (always a winning coalition). Thus, the total number of winning coalitions is 2^{N-1}.

77. (a) According to Table 2-11, the Banzhaf index of California in the Electoral College is $\beta_{California} = 0.114$.

The relative voting weight of California is $w_{California} = \dfrac{55}{538} \approx 0.1022$. The relative Banzhaf voting power

for California is then $\pi_{California} = \dfrac{0.114}{0.1022} \approx 1.115$.

(b) $\pi_{H1} = \dfrac{1/3}{31/115} = \dfrac{115}{93}$, $\pi_{H2} = \dfrac{1/3}{31/115} = \dfrac{115}{93}$, $\pi_{OB} = \dfrac{1/3}{28/115} = \dfrac{115}{84}$, $\pi_{NH} = \dfrac{0}{21/115} = 0$,

$\pi_{LB} = \dfrac{0}{2/115} = 0$, $\pi_{GC} = \dfrac{0}{2/115} = 0$.

79. (a) [4: 3,1,1,1]; P_1 is pivotal in every sequential coalition except the six sequential coalitions in which P_1 is the first member. So, the Shapley-Shubik power index of P_1 is $(24 - 6)/24 = 18/24 = 3/4$. P_2, P_3, and P_4 share power equally and have Shapley-Shubik power index of $1/12$.

(b) In a weighted voting system with 4 players there are 24 sequential coalitions — each player is the first member in exactly 6 sequential coalitions. The only way the first member of a coalition can be pivotal is if she is a dictator. Consequently, if a player is not a dictator, she can be pivotal in at most $24 - 6 = 18$ sequential coalitions and so that player's Shapley-Shubik power index can be at most $18/24 = 3/4$.

(c) In a weighted voting system with N players there are $N!$ sequential coalitions — each player is the first member in exactly $(N–1)!$ sequential coalitions. The only way the first member of a coalition can be pivotal is if she is a dictator. Consequently, if a player is not a dictator, she can be pivotal in at most $N! - (N–1)! = (N–1)!(N–1)$ sequential coalitions and so that player's Shapley-Shubik power index can be at most $(N–1)!(N–1)/N! = (N–1)/N$.

(d) [N: $N–1$, 1, 1, 1,…, 1]; Here N is the number of players as well as the quota. P_1 is pivotal in every sequential coalition except for the $(N–1)!$ sequential coalitions in which P_1 is the first member. So the Shapley-Shubik power index of P_1 is $[N! - (N–1)!]/N! = (N–1)/N$.

Chapter 3

WALKING

3.1   Fair-Division Games

1. **(a)** When three players are dividing assets, a fair share to any player is one worth at least 1/3 of the total. So any share worth more than $33\frac{1}{3}$% would be considered a fair share. In this case, s_2 and s_3 (since, in Henry's eyes, they are worth 40% and 35% respectively) are fair.

   (b) s_2 and s_3 (35% and 37% are each valued by Tom as at least $33\frac{1}{3}$%)

   (c) s_1, s_2 and s_3

   (d) Since Fred is the only player to consider s_1 to be fair, any fair division of the assets must allocate s_1 to Fred. Since Henry and Tom each consider both s_2 and s_3 to be fair shares, there are exactly two possible divisions:
   (1) Henry gets s_2; Tom gets s_3; Fred gets s_1.
   (2) Henry gets s_3; Tom gets s_2; Fred gets s_1.

   (e) Since Henry values s_2 more than s_3 and Tom values s_3 more than s_2, the best division is
   (1) Henry gets s_2; Tom gets s_3; Fred gets s_1.

3. **(a)** When four players are dividing assets, a fair share to any player is one worth at least 1/4 or 25% of the total. In this case, s_2 and s_3 (since they are worth 26% and 28% respectively in Angie's eyes) are fair.

   (b) s_1, s_2 and s_4

   (c) s_2 and s_3

   (d) s_1, s_2, s_3 and s_4

   (e) The following table can help in determining possible fair divisions. A check denotes a fair share.

	s_1	s_2	s_3	s_4
Angie		✓	✓	
Bev	✓	✓		✓
Ceci		✓	✓	
Dina	✓	✓	✓	✓

   Suppose we allocate share s_2 to Angie. Since Ceci only considers s_2 and s_3 to be fair shares, Ceci must be allocated s_3. Since Dina finds all shares fair, this leaves two choices for the share we may allocate to Bev - s_1 or s_4. These are the first two of our fair divisions.
   (1) Angie gets s_2; Bev gets s_1; Ceci gets s_3; Dina gets s_4.
   (2) Angie gets s_2; Bev gets s_4; Ceci gets s_3; Dina gets s_1.

Suppose instead that we allocate s_3 to Angie. Since Ceci only considers s_2 and s_3 to be fair shares, Ceci must be allocated s_2. Since Dina finds all shares fair, this leaves two choices for the share we may allocate to Bev - s_1 or s_4. This gives our other two fair divisions.

(3) Angie gets s_3; Bev gets s_1; Ceci gets s_2; Dina gets s_4.

(4) Angie gets s_3; Bev gets s_4; Ceci gets s_2; Dina gets s_1.

5. (a) When four players are dividing assets, a fair share to any player is one worth at least 1/4 of the total values of the assets in their eyes (the total value for each player, in dollars, may differ). In this case, Allen values this cake at $\$4 + \$5 + \$6 + \$5 = \$20$. So fair shares to Allen are those worth at least $\$5.00$. These are s_2, s_3, and s_4.

(b) Brady sees the cake as being worth a total of $\$16$. So fair shares to Brady are those worth at least $\$4.00$. He sees s_3 and s_4 as fair.

(c) Cody sees the cake as being worth a total of $\$18$. So fair shares to Cody are those worth at least $\$4.50$. He sees s_1 and s_2 as fair.

(d) Diane sees the cake as being worth a total of $\$20$. So fair shares to Diane are those worth at least $\$5.00$. She sees s_1 and s_4 as fair.

(e) The following table can help in determining possible fair divisions. A check denotes a fair share.

	s_1	s_2	s_3	s_4
Allen		✓	✓	✓
Brady			✓	✓
Cody	✓	✓		
Diane	✓			✓

Suppose we allocate share s_2 to Allen. Since Cody only considers s_1 and s_2 to be fair shares, Cody must be allocated s_1. But then since Diane only considers s_1 or s_4 to be fair, Diane must be allocated s_4. This leaves Brady with s_3. This gives the first of our fair divisions.

(1) Allen gets s_2; Brady gets s_3; Cody gets s_1; Diane gets s_4.

Suppose now that we allocate s_3 to Allen. Since Brady only considers s_3 and s_4 to be fair shares, Brady must be allocated s_4. But then since Diane only considers s_1 or s_4 to be fair, Diane must be allocated s_1. This leaves Cody with s_2. This gives the second of our fair divisions.

(2) Allen gets s_3; Brady gets s_4; Cody gets s_2; Diane gets s_1.

Lastly, suppose that we allocate s_4 to Allen. Since Brady only considers s_3 and s_4 to be fair shares, Brady must be allocated s_3. Also, since Diane only considers s_1 or s_4 to be fair, Diane must be allocated s_1. This leaves Cody with s_2. This gives the third and last of our fair divisions.

(3) Allen gets s_4; Brady gets s_3; Cody gets s_2; Diane gets s_1.

7. (a) Let x denote the value of slices s_2 and s_3 to Adams. Then,

$$(x + \$40,000) + x + x + (x + \$60,000) = \$400,000 \text{ and so } x = \$75,000.$$ Since a fair share is worth $\$100,000$, Adams considers slices s_1 ($\$115,000$) and s_4 ($\$135,000$) to be fair shares.

(b) Let x denote the value of slice s_3 to Benson. Then, since s_4 is worth \$8000 more than that and the sum of these two is worth $(0.40)\$400,000 = \$160,000$, it follows that $x + (x + \$8,000) = \$160,000$ and so $x =$ \$76,000. So, s_3 is worth \$76,000 and s_4 is worth \$84,000.

Now let y denote the value of slice s_2 to Benson. The value of s_1 is then $y + \$40,000$. So, $(y + \$40,000) + y + \$76,000 + \$84,000 = \$400,000$ and $y = \$100,000$. Since a fair share is \$100,000, Benson considers slices s_1 (\$140,000) and s_2 (\$100,000) to be fair shares.

(c) Let x denote the value of slice s_4 to Cagle. Then, s_3 is worth $2x$ and s_1 is worth $x + \$20,000$. Since s_2 is worth \$40,000 less than s_1, the value of s_2 is x - \$20,000. Hence, $(x + \$20,000) + (x - \$20,000) + 2x + x = \$400,000$ and $x = \$80,000$. Since a fair share is \$100,000, Cagle considers slices s_1 (\$100,000) and s_3 (\$160,000) to be fair shares.

(d) Let x denote the value of slices s_2 and s_3 to Duncan. Then, $(x + \$4,000) + x + x = (0.70)(\$400,000) =$ \$280,000 so that $x = \$92,000$. Since a fair share is \$100,000, Duncan only considers slice s_4 (\$120,000) to be a fair share.

(e) Duncan must receive s_4. Then, Adams must receive s_1. After that, Benson must receive s_2 leaving Cagle with s_3.

9. (a) Let C = the value of the chocolate half (in Angelina's eyes)
H = the value of the hibiscus half (in Angelina's eyes)
It is known that $H = 2C$ (Angelina likes the hibiscus half twice as much as the chocolate half) and $C + H = \$72$ (the entire cake is worth \$72). Substituting,
$C + 2C = \$72$
$3C = \$72$
$C = \$24$
So, the chocolate part is valued by Angelina at \$24 and the hibiscus part is valued by Angelina at \$72-\$24
= \$48. The hibiscus wedge is then valued as $\dfrac{60^\circ}{180^\circ} \times \$48 = \dfrac{1}{3} \times \$48 = \$16$ by Angelina.

(b) By (a), the chocolate piece shown is worth $\dfrac{45^\circ}{180^\circ} \times \$24 = \dfrac{1}{4} \times \$24 = \$6$ in Angelina's eyes.

(c) By (a), the sliver of hibiscus cake shown is worth $\dfrac{12^\circ}{180^\circ} \times \$48 = \$3.20$ in Angelina's eyes.

11. (a) Let H = the value of the hibiscus half (in Brad's eyes)
C = the value of the chocolate half (in Brad's eyes)
It is known that $C = 4H$ (Brad likes the chocolate half four times as much as the hibiscus half) and $C + H = \$72$ (the entire cake is worth \$72). Substituting,
$4H + H = \$72$
$5H = \$72$
$H = \$14.40$
So, the hibiscus half is valued by Brad at \$14.40 and the chocolate half is valued by Brad at \$72-\$14.40 =
\$57.60. The hibiscus wedge is then valued as $\dfrac{60^\circ}{180^\circ} \times \$14.40 = \dfrac{1}{3} \times \$14.40 = \$4.80$ by Brad.

(b) By (a), the chocolate piece shown is worth $\dfrac{45^\circ}{180^\circ} \times \$57.60 = \dfrac{1}{4} \times \$57.60 = \$14.40$ in Brad's eyes.

(c) By (a), the sliver of hibiscus cake shown is worth $\dfrac{12^\circ}{180^\circ} \times \$14.40 = \$0.96$ in Brad's eyes.

13. (a) Let C = the value of the chocolate part (in Karla's eyes)

S = the value of the strawberry part (in Karla's eyes)

V = the value of the vanilla part (in Karla's eyes)

It is known that $S = 2V$ (Karla likes strawberry twice as much as vanilla), $C = 3V$ (Karla likes chocolate three times as much as vanilla), and $C + S + V = \$30$ (the entire cake is worth \$30). Substituting,

$$3V + 2V + V = \$30$$
$$6V = \$30$$
$$V = \$5$$
$$S = 2 \times (\$5) = \$10$$
$$C = 3 \times (\$5) = \$15$$

That is, the vanilla part is worth \$5, the strawberry part \$10, and the chocolate part \$15 in Karla's eyes. The value of each slice is then found as follows.

$$s_1 : \dfrac{60^\circ}{120^\circ} \times \$5 = \$2.50 \; ; \; s_2 : \dfrac{30^\circ}{120^\circ} \times \$5 + \dfrac{30^\circ}{120^\circ} \times \$10 = \$1.25 + \$2.50 = \$3.75 \; ;$$

$$s_3 : \dfrac{60^\circ}{120^\circ} \times \$10 = \$5.00 \; ; \; s_4 : \dfrac{30^\circ}{120^\circ} \times \$10 + \dfrac{30^\circ}{120^\circ} \times \$15 = \$2.50 + \$3.75 = \$6.25 \; ;$$

$$s_5 : \dfrac{60^\circ}{120^\circ} \times \$15 = \$7.50 \; ; \; s_6 : \dfrac{30^\circ}{120^\circ} \times \$15 + \dfrac{30^\circ}{120^\circ} \times \$5 = \$3.75 + \$1.25 = \$5.00$$

(b) A fair share would need to be worth at least \$30/6 = \$5.00. So, slices s_3, s_4, s_5, and s_6 are each fair shares.

3.2   The Divider–Chooser Method

15. (a) Since Jared likes meatball subs three times as much as vegetarian subs, he values each of the pieces [0,6], [6,8], [8,10], and [10,12] the same. So, with one cut Jared would divide the sandwich into parts $s_1 : [0,8]$ and $s_2 : [8,12]$. That is, one piece containing the entire vegetarian half and one-third of the meatball half (s_1) and one piece consisting of two-thirds of the meatball half (s_2).

(b) Karla would clearly choose piece s_1 since she only values vegetarian. This piece contains the entire value of the sandwich to Karla (\$8.00).

17. (a) Let x denote the value that Martha places on one inch of turkey (in dollars). Then, Martha places that same value on one inch of roast beef. However, she places a value of $2x$ on each inch of ham. Since the total value of the sandwich is \$9, we have $8(2x) + 12(x) + 8(x) = \$9$. That is, $36x = \$9$ or $x = \$0.25$. Each inch of ham is worth \$0.50 to Martha and each inch of turkey and roast beef is worth \$0.25. To have two pieces each worth \$4.50, she would cut the sandwich into pieces $s_1 : [0,10]$ and $s_2 : [10,28]$.

(b) Let y denote the value that Nick places on one inch of turkey (in dollars). Then, Nick places that same value on one inch of ham. However, he places a value of $2y$ on each inch of roast beef. Since the total value of the sandwich is \$9, we have $8(y) + 12(y) + 8(2y) = \$9$. That is, $36y = \$9$ or $y = \$0.25$. Each inch of roast beef is worth \$0.50 to Nick and each inch of turkey and ham is worth \$0.25. Nick would clearly take piece $s_2 : [10,28]$ which, in his eyes, has a value of $10(\$0.25) + 8(\$0.50) = \$6.50$.

19. (a) Yes; David values chocolate and vanilla equally and values each twice as much as strawberry. Paula would choose s_2 since she is allergic to chocolate (s_2 contains ¾ of the value of the entire cake in her eyes).

(b) No; David values s_2 more than he values s_1. In fact, he values it 1.5 times more. That is, he values at 60% and s_2 and s_1 at 40%. Such a division is not rational since there is a risk that Paula, whom he knows nothing about, would choose that piece.

(c) Yes; Based on David's value system, the chocolate and vanilla parts are each worth twice as much as the strawberry part. So, if the entire cake is worth \$5 (hypothetically), then, the chocolate and vanilla parts would each be worth \$2 and the strawberry part would be valued at \$1. In that case, s_1 is worth

$$\frac{90°}{120°}(\$2) + \frac{60°}{120°}(\$2) = \$2.50,$$ or half the full value of the cake. Paula would choose the piece with the least chocolate (piece s_2).

3.3 The Lone-Divider Method

21. (a) D must be given s_1. But C_1 and C_2 are happy with either of the other shares. So, the following are two fair divisions of the land.
$C_1 : s_2, \ C_2 : s_3, \ D : s_1$;
$C_1 : s_3, \ C_2 : s_2, \ D : s_1$

(b) If C_1 is assigned s_2, then C_2 can be assigned either of the other pieces. However, if C_1 is assigned s_3, then C_2 must be assigned s_1. So, the following are three fair divisions.
$C_1 : s_2, \ C_2 : s_1, \ D : s_3$;
$C_1 : s_2, \ C_2 : s_3, \ D : s_1$;
$C_1 : s_3, \ C_2 : s_1, \ D : s_2$

23. (a) First, C_1 must receive s_2. Then, C_3 must receive s_3 (the only other share C_3 considers fair). So, C_2 receives s_1 and D receives s_4.

(b) If C_1 receives s_2, then C_3 must receive s_1. But then C_2 must receives s_3 leaving D with s_4. If, on the other hand, C_1 receives s_3, then C_2 must receive s_1. But then C_3 must receives s_2 leaving D with s_4. Hence, two fair divisions are as follows.
$C_1 : s_2, \ C_2 : s_3, \ C_3 : s_1, \ D : s_4$;
$C_1 : s_3, \ C_2 : s_1, \ C_3 : s_2, \ D : s_4$

(c) C_1 must receive s_2. If C_2 receives s_1, then C_3 must receive s_4 leaving D with s_3. On the other hand, if C_2 receives s_3, then C_3 could receive either s_1 or s_4 leaving D with the remaining piece. Three fair divisions follow below.
$C_1 : s_2, \ C_2 : s_1, \ C_3 : s_4, \ D : s_3$;
$C_1 : s_2, \ C_2 : s_3, \ C_3 : s_1, \ D : s_4$;
$C_1 : s_2, \ C_2 : s_3, \ C_3 : s_4, \ D : s_1$

25. (a) Since Tim valued each piece at 25% (and was the only player to do so), he must have been the divider.

(b) Based on the percentages listed, the following would be the bids made.

   Mark : $\{s_2, s_3\}$, *Maia* : $\{s_3, s_4\}$, *Kelly* : $\{s_4\}$.

   So, a fair division of the cake must have Kelly with slice s_4. But then Maia must get slice s_3. That would require Mark to receive slice s_2. This leaves the divider, Tim, with slice s_1.

27. (a) Fair; none of the bidders considered s_3 as being worth ¼ of the land. So, these three shares must be worth at least ¾ of the land in each of their eyes. If they each wind up with at least 1/3 of the newly combined piece, that would constitute at least 1/3 of 3/4 (i.e. ¼) of the value of the original plot of land.

   (b) Not fair; $\{s_2, s_3, s_4\}$ may be worth less than ¾ of the value of the original plot of land to C_2 and C_3. Even if they wind up with 1/3 of the newly combined piece, it may not be a fair share.

   (c) Fair; see (a).

   (d) Not fair; $\{s_1, s_4\}$ may be worth less than ½ of the value of the original plot of land to C_2 and C_3. That is, s_2 may be of a lot of value to C_2 and C_3.

29. (a) $C_1 : s_2$, $C_2 : s_4$, $C_3 : s_5$, $C_4 : s_3$, $D : s_1$;

   $C_1 : s_4$, $C_2 : s_2$, $C_3 : s_5$, $C_4 : s_3$, $D : s_1$;

   If C_1 is to receive s_2, then C_2 must receive s_4 (and conversely). So, C_4 must receive s_3. It follows that C_3 must receive s_5. This leaves D with s_1.

   (b) $C_1 : s_2$, $C_2 : s_4$, $C_3 : s_5$, $C_4 : s_3$, $D : s_1$;

   If C_1 is to receive s_2, then C_2 must receive s_4. So, C_4 must receive s_3. It follows that C_3 must receive s_5. This leaves D with s_1.

31. (a) The Divider was Gong since that is the only player that could possibly value each piece equally.

   (b) To determine the bids placed, each player's bids should add up to $480,000. So, for example, Egan would bid $480,000 - $80,000 - $85,000 - $195,000 = $120,000 on parcel s_3. Further, one player must bid the same on each parcel (the divider). The following table shows the value of the four parcels in the eyes of each partner.

	s_1	s_2	s_3	s_4
Egan	$80,000	$85,000	$120,000	$195,000
Fine	$125,000	$100,000	$135,000	$120,000
Gong	$120,000	$120,000	$120,000	$120,000
Hart	$95,000	$100,000	$175,000	$110,000

   For a player to bid on a parcel, it would need to be worth at least $480,000/4 = $120,000. Based on this table, the choosers would make the following declarations:
   Egan: $\{s_3, s_4\}$; Fine: $\{s_1, s_3, s_4\}$; Hart: $\{s_3\}$.

   (c) One possible fair division of the land is Egan: s_4; Fine: s_1, Gong: s_2; Hart: s_3. In fact, this is the only possible division of the land since Hart must receive s_3, so that Egan in turn must receive s_4. It follows that Fine must receive s_1 so that Gong (the divider) winds up with s_2.

33. (a) $s_1 : [0,4]$; $s_2 : [4, 8]$; $s_3 : [8, 12]$

(b) Karla values [0,2], [2,4], and [4,6] equally. She places no value on the segment [6,12]. So, Karla would consider any piece that contains [0,2] or [2,4] or [4,6] to be a fair share. It follows that Karla views s_1 and s_2 as fair shares.

(c) Suppose that Lori places a value of $1 (or 1 cent, or one peso, or whatever) on one inch of vegetarian sub. To her then the entire sub is worth $6 + $12 = $18. Also, she values s_1 at $4, s_2 at $6, and s_3 at $8. Since a fair share must be worth $6 to Lori, it follows that s_2 and s_3 are fair shares.

(d) Three possible fair divisions are
Jared: s_2, Karla: s_1, Lori: s_3 ;
Jared: s_1, Karla: s_2, Lori: s_3 ;
Jared: s_3, Karla: s_1, Lori: s_2

3.4   The Lone-Chooser Method

35. (a) Angela sees s_1 as being worth $18. In her second division, she will create three $6 pieces.

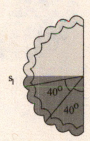

Since the strawberry quarter at the lower left is worth $13.50 to Angela, the size of the central angle in one pie slice she will cut will be determined by

$$\frac{x}{90°}(\$13.50) = \$6 \text{ so that } x = 40°.$$

In fact, Angela cuts two 40° pieces from the strawberry, and the remaining strawberry plus the vanilla makes up the third piece. The pieces can be described as 40° strawberry, 40° strawberry, and 90° vanilla-10° strawberry.

(b) Boris sees s_2 as being worth $15. In his second division, he will create three $5 pieces.

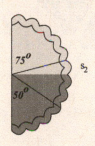

Since the strawberry quarter at the lower right is worth $9.00 to Boris, the size of the central angle in one pie slice he will cut will be determined by

$$\frac{x}{90°}(\$9.00) = \$5.00 \text{ so that } x = 50°. \text{ So Boris cuts one } 50° \text{ piece from the}$$

strawberry part of the cake. Also, since the vanilla quarter at the upper right is

worth $6.00 to Boris, we see that $\frac{x}{90°}(\$6.00) = \$5.00$ tells us that $x = 75°$ is

the size of the central angle in a second pie slice (vanilla this time) Boris makes. The remaining 15° piece of vanilla plus the 40° piece of strawberry makes up the third piece. The pieces can be described as 75° vanilla, 15° vanilla-40° strawberry, 50° strawberry.

(c) Since Carlos values vanilla twice as much as strawberry, Carlos would clearly take the vanilla part of Angela's division (the 90° vanilla-10° strawberry wedge). He would also want to take the most vanilla that he could from Boris (the 75° vanilla wedge). One possible fair division is Angela: 80° strawberry, Boris: 15° vanilla-90° strawberry, and Carlos: 165° vanilla-10° strawberry.

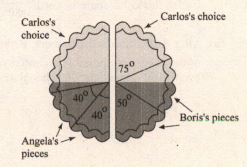

(d) Since she receives two of her $6.00 pieces, the value of Angela's final share (in Angela's eyes) is $12.00. Similarly, the value of Boris' final share (in Boris' eyes) is $10.00. The value of Carlos' final share (in Carlos' eyes) is $\frac{90°}{90°}($12$) + \frac{10°}{90°}($6$) + \frac{75°}{90°}($12$) \approx 22.67.

37. First, note that the value of s_1 (in Angela's eyes) is $\frac{120°}{180°} \times $27 = 18. She values s_2 at $18 too.

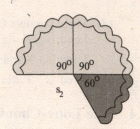

(a) Boris chooses s_2 because he views it as being worth

$12 + $\frac{60°}{180°} \times $18 = $12 + $6 = 18 (s_1 is only worth $12.00 to him).

Boris views s_2 as being worth $18 and divides it into three pieces each worth $6.00. Thus, one possible second division of s_2 by Boris is 90° vanilla, 90° vanilla, 60° strawberry.

(b) Angela divides s_1 evenly (40° strawberry, 40° strawberry, 40° strawberry).

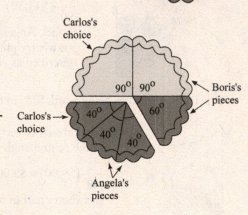

(c) Carlos will select any one of Angela's pieces since they are identical in value to him. He will also select one of Boris's vanilla wedges since he values vanilla twice as much as he values strawberry.

One possible fair division is Angela: 80° strawberry, Boris: 90° vanilla-60° strawberry, and Carlos: 90° vanilla-40° strawberry.

(d) Angela thinks her share is worth $\frac{80°}{180°} \times $27 = 12.00.

Boris thinks his share is worth $\frac{90°}{180°} \times $12 + \frac{60°}{180°} \times $18 = 12.00.

Carlos thinks his share is worth $\frac{40°}{180°} \times $12 + \frac{90°}{180°} \times $24 = 14.67.

39. (a) After Arthur makes the first cut, Brian chooses s_1. To him, it is worth 100% of the cake. One possible second division by Brian is to divide the piece evenly (60° chocolate, 30° chocolate-30° strawberry, 60° strawberry).

(b) Arthur places all of the value on the orange half of s_2, so he divides the orange evenly. The second division by Arthur can be described as 30° orange, 30° orange, 30° orange-90° vanilla.

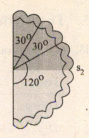

(c) Since Carl likes chocolate and vanilla, one possible fair division is Arthur: 60° orange, Brian: 90° strawberry-30° chocolate, and Carl: 90° vanilla-30° orange-60° chocolate.

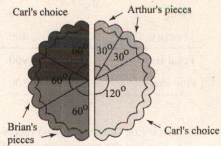

(d) Arthur thinks his share is worth $33\frac{1}{3}$%. Brian thinks his share is worth $66\frac{2}{3}$% (remember, he is getting 2/3 of s_1 which he values at 100%). Carl thinks his share is worth

$$\frac{60°}{90°} \times 50\% + \frac{90°}{90°} \times 50\% + \frac{30°}{90°} \times 0\% = 83\frac{1}{3}\%.$$

41. (a) Ignoring the meatball part (having no value to her), Karla would cut the sandwich right down the middle of the vegetarian part. That is, s_1: [0,3], s_2:[3,12].

(b) Since Jared likes meatball and vegetarian the same, he will pick the larger half (by size), namely s_2. He will then divide his choice (s_2:[3,12]) equally as J_1:[3,6], J_2:[6,9], J_3:[9,12].

(c) Karla's subdivision is just as simple as Jared's – she likes all parts of s_1 equally. So her division can be described as K_1:[0,1], K_2:[1,2], K_3:[2,3].

(d) Since Lori likes meatball twice as much as vegetarian, she will select any of Karla's identical pieces (say, K_3:[2,3] is selected). She will also select either J_2 or J_3 from Jared (say, J_3:[9,12]). So, one fair division would be described by Jared: [3,9], Karla: [0,2], Lori: [2,3] and [9,12]. In this case, Jared ends up with half of the sandwich (in size) worth 50% of the value of the whole sandwich to him. Karla ends up with 1/3 of the vegetarian half of the sandwich worth $33\frac{1}{3}$% of the value of the whole sandwich to her. Lori values the vegetarian half at 1/3 and the meatball part at 2/3 so that her piece is valued at

$$\frac{1}{6} \times \frac{1}{3} + \frac{1}{2} \times \frac{2}{3} = \frac{7}{18} = 38\frac{8}{9}\%.$$

3.5   The Method of Sealed Bids

43. (a) Ana and Belle's fair share is $300. Chloe's fair share is $400.

Item	Ana	Belle	Chloe
Dresser	150	**300**	275
Desk	**180**	150	165
Vanity	170	200	**260**
Tapestry	400	250	**500**
Total Bids	900	900	1200
Fair share	300	300	400

(b) In the first settlement, Ana receives the desk and $120 in cash; Belle receives the dresser; Chloe gets the vanity and the tapestry and pays $360.

	Ana	Belle	Chloe
Total Bids	900	900	1200
Fair share	300	300	400
Value of items received	180	300	760
Prelim cash settlement	120	0	−360

(c) $240; $360 was paid in and $120 was paid out in the first settlement.

(d) Ana, Belle, and Chloe split the $240 surplus cash three ways. This adds $80 cash to the first settlement. In the final settlement, Ana gets the desk and receives $200 in cash; Belle gets the dresser and receives $80; Chloe gets the vanity and the tapestry and pays $280.

Item	Ana	Belle	Chloe
Prelim cash settlement	120	0	−360
Share of surplus	80	80	80
Final cash settlement	200	80	−280

45. (a) Each player's fair share is the sum of their bids divided by 5.

Item	A	B	C	D	E
Item 1	$352	$295	**$395**	$368	$324
Item 2	$98	$102	$98	$95	**$105**
Item 3	$460	$449	**$510**	$501	$476
Item 4	**$852**	$825	$832	$817	$843
Item 5	**$513**	$501	$505	$505	$491
Item 6	$725	$738	$750	$744	**$761**
Total Bids	$3000	$2910	$3090	$3030	$3000
Fair Share	$600	$582	$618	$606	$600

(b) After the first settlement, *A* ends up with items 4 and 5 and pays $765, *B* gets $582, *C* receives items 1 and 3 and pays $287, *D* gets $606, and *E* receives items 2 and 6 and pays $266.

	A	B	C	D	E
Total Bids	3000	2910	3090	3030	3000
Fair Share	600	582	618	606	600
Value of items rec'd	1365	0	905	0	866
Prelim. cash	−765	582	−287	606	−266

(c) There is a surplus of $765 - $582 + $287 - $606 + $266 = $130.

(d) *A*, *B*, *C*, *D*, and *E* split the $130 surplus 5 ways ($26 each). In the final settlement, *A* ends up with items 4 and 5 and pays $739; *B* ends up with $608; *C* ends up with items 1 and 3 and pays $261; *D* ends up with $632; *E* ends up with items 2 and 6 and pays $240.

47. Anne gets $75,000 and Chia gets $80,000.

Item	Anne	Bette	Chia	
Partnership	$210,000	$240,000	$225,000	
Total Bids	$210,000	$240,000	$225,000	
Fair Share	$70,000	$80,000	$75,000	**Total**
Value of items received	0	$240,000	0	**Surplus**
Prelim. cash settlement	$70,000	−$160,000	$75,000	**$15,000**
Share of surplus	$5,000	$5,000	$5,000	
Final cash settlement	$75,000	−$155,000	$80,000	

49. In this case, the low bid is awarded for each chore. The total bids here represent the amount of money they are willing to be paid to do all four chores. Ali does chore 4 and pays Caren $15, Briana does chore 2 and receives no cash, and Caren does chores 1 and 3 and receives $15 from Ali.

Item	Ali	Briana	Caren	
Chore 1	$65	$70	**$55**	
Chore 2	$100	**$85**	$95	
Chore 3	$60	$50	**$45**	
Chore 4	**$75**	$80	$90	
Total Bids	$300	$285	$285	
Fair Share	$100	$95	$95	**Total**
Value of chores done	$75	$85	$100	**Surplus**
Prelim. cash settlement	-$25	-$10	$5	**$30**
Share of surplus	$10	$10	$10	
Final cash settlement	$-15	$0	$15	

3.6 The Method of Markers

51. (a) Using the first first marker, B gets items 1, 2, 3; Then, using the first second marker, C gets items 5, 6, 7; Finally, A gets items 10, 11, 12, 13.

(b) B gets items 1, 2, 3.

(c) C gets items 5, 6, 7.

(d) Items 4, 8, and 9 are left over.

53. (a) Using the first first marker, A gets items 1, 2; Then, using the first second marker, C gets items 4, 5, 6, 7; Finally, B gets items 10, 11, 12.

(b) B gets items 10, 11, 12

(c) C gets items 4, 5, 6, 7

(d) Items 3, 8, and 9 are left over.

55. (a) First C gets items 1, 2, 3; Then, E gets items 5, 6, 7, 8; Then, D gets items 11, 12, 13; Then, B gets items 15, 16, 17; Finally, A gets items 19, 20.

(b) Items 4, 9, 10, 14, and 18 are left over.

57. (a) Quintin thinks the total value is $3 \times \$12 + 6 \times \$7 + 6 \times \$4 + 3 \times \$6 = \$120$, so to him a fair share is worth $30. Ramon thinks the total value is $3 \times \$9 + 6 \times \$5 + 6 \times \$5 + 3 \times \$11 = \$120$, so to him a fair share is worth $30. Stephone thinks the total value is $3 \times \$8 + 6 \times \$7 + 6 \times \$6 + 3 \times \$14 = \$144$, so to him a fair share is worth $36. Tim thinks the total value is $3 \times \$5 + 6 \times \$4 + 6 \times \$4 + 3 \times \$7 = \$84$, so to him a fair share is worth $21.

They would place their markers as shown below.

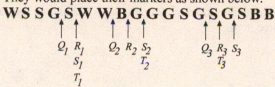

(b)

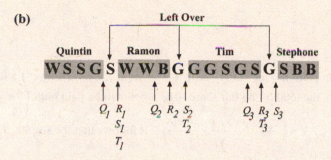

59. **(a)** Ana places her markers between the different types of candy. Belle places her markers between each of the Minto bars. Chloe places her markers thinking that each Choko and Minto bar is worth $1 and each Frooto is worth $2. Since Chloe would then value the candy to be worth $12 in total, she would place markers after each $4 worth of candy.

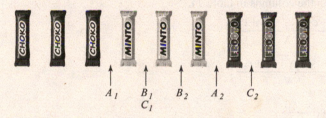

(b) Ana gets three Choko bars, Belle gets one Minto bar, and Chloe gets two Frooto bars. Two Minto bars and one Frooto bar are left over.

(c) Belle would select one of the Minto bars, Chloe would then select the Frooto bar, and finally Ana would be left with the last Minto bar.

JOGGING

61. **(a)** C gets to choose one of the three pieces. One of the three must be worth more than 1/3 of S.

(b) If C chooses either s_{21} or s_{22}, D_1 gets to choose s_1 (a fair share to him). If C chooses s_1, D_1 gets to choose between s_{21} and s_{22}. Together they are worth 2/3 of S, so one of the two must be worth at least 1/3 of S.

(c) Suppose that in D_2's opinion, the first cut by D_1 splits S into a 40%-60% split. In this case, the value of the three pieces to D_2 is $s_1 = 40\%$, $s_{21} = 30\%$, and $s_{22} = 30\%$. If C or D_1 choose s_1, D_2 will not get a fair share.

63. In the first settlement A receives $\$x/2$; B receives the partnership and pays $\$y/2$. The surplus is

$$\frac{y}{2} - \frac{x}{2} = \frac{y-x}{2} \text{ dollars. } B \text{ must pay } A\text{'s original fair share plus half of the surplus, or (in dollars)}$$

$$\frac{x}{2} + \frac{1}{2}\left(\frac{y-x}{2}\right) = \frac{x}{2} + \left(\frac{y-x}{4}\right) = \frac{x}{2} + \frac{y}{4} - \frac{x}{4} = \frac{x}{4} + \frac{y}{4} = \frac{x+y}{4}.$$

65. *B* receives the strawberry part. *C* will receive 5/6 of the chocolate part. *A* receives the vanilla part and 1/6 of the chocolate part.

67. (a) Two, three, or four choosers bidding on the same item; three or four choosers bidding on the same two items; four choosers bidding on the same three items.

(b) $1 + 2 + 3 + 4 = 10$

(c) $1 + 2 + 3 + \ldots + (N-2) = \dfrac{(N-1)(N-2)}{2}$

RUNNING

69. (a) Based on the M winning bids, the amount paid into the "kitty" is $w_1 + w_2 + w_3 + \ldots + w_M$. Player j's fair share (based on their bids) is c_j / N. So, the total of the fair shares (the amount to be paid out of the "kitty") is $c_1 / N + c_2 / N + c_3 / N + \ldots + c_N / N = \left(\dfrac{c_1 + c_2 + c_3 + \ldots + c_N}{N} \right)$. It follows that the surplus S is given by $S = \left(w_1 + w_2 + w_3 + \ldots + w_M \right) - \left(\dfrac{c_1 + c_2 + c_3 + \ldots + c_N}{N} \right)$.

(b) The total value of *all* bids is $c_1 + c_2 + c_3 + \ldots + c_N$. But this is the same as $r_1 + r_2 + r_3 + \ldots + r_M$ (here we are just summing the rows rather than the columns in table *T*). So,

$$S = \left(w_1 + w_2 + w_3 + \ldots + w_M \right) - \left(\frac{c_1 + c_2 + c_3 + \ldots + c_N}{N} \right)$$

$$= \left(w_1 + w_2 + w_3 + \ldots + w_M \right) - \left(\frac{r_1 + r_2 + r_3 + \ldots + r_M}{N} \right)$$

$$= \left(w_1 - \frac{r_1}{N} \right) + \left(w_2 - \frac{r_2}{N} \right) + \ldots + \left(w_M - \frac{r_M}{N} \right)$$

(c) Since the winning bid on item k is always greater than or equal to the average bid on that item, $w_k - \dfrac{r_k}{N} \geq 0$. Since this is true for every item, $S \geq 0$.

(d) The surplus $S = 0$ exactly when the winning bid on each item k is the same as the average bid on that item. That is, when all players bid the exact same amount on each item.

71. After the division, suppose that the divisors end up with shares $s_1, s_2, \ldots, s_{N-1}$ respectively. The sum of the values of these shares in the chooser *C*'s eyes will be 1. Call these values $x_1, x_2, \ldots, x_{N-1}$ so that $x_1 + x_2 + \ldots + x_{N-1} = 1$. After the subdivision by each divider into N subshares, the chooser will select shares worth at least $\dfrac{1}{N} \cdot x_1, \dfrac{1}{N} \cdot x_2, \ldots, \dfrac{1}{N} \cdot x_{N-1}$. So, the total value of her selection will be worth at least

$\dfrac{1}{N} \cdot x_1 + \dfrac{1}{N} \cdot x_2 + \ldots + \dfrac{1}{N} \cdot x_{N-1} = \dfrac{1}{N} \left(x_1 + x_2 + \ldots + x_{N-1} \right) = \dfrac{1}{N}$. The dividers, on the other hand, will end up with a collection of subshares worth exactly $\dfrac{(N-1)}{N}$ of at least $\dfrac{1}{N-1}$ of the value of the entire cake. That is, worth at least $1/N$ of the value of the cake.

Chapter 4

WALKING

4.1. Apportionment Problems and Apportionment Methods

1. (a) To determine the standard divisor, we find the total population of the states and divide by the number of available seats.

$$\text{Standard divisor} = \frac{3,310,000 + 2,670,000 + 1,330,000 + 690,000}{160} = 50,000$$

(b) To determine each state's standard quota, we divide that state's population by the standard divisor found in part (a).

Apure: $\dfrac{3,310,000}{50,000} = 66.2$; Barinas: $\dfrac{2,670,000}{50,000} = 53.4$;

Carabobo: $\dfrac{1,330,000}{50,000} = 26.6$; Dolores: $\dfrac{690,000}{50,000} = 13.8$

3. (a) Standard divisor $= \dfrac{45,300 + 31,070 + 20,490 + 14,160 + 10,260 + 8,720}{125} = 1040$

(b) The "states" in any apportionment problem are the entities that will have "seats" assigned to them according to a share rule. In this case, the states are the six bus routes and the seats are the 125 buses. The standard divisor of 1,040 represents the average number of passengers per bus per day.

(c) A: $\dfrac{45,300}{1040} = 43.558$; B: $\dfrac{31,070}{1040} = 29.875$; C: $\dfrac{20,490}{1040} = 19.702$;

D: $\dfrac{14,160}{1040} = 13.615$; E: $\dfrac{10,260}{1040} = 9.865$; F: $\dfrac{8,720}{1040} = 8.385$

5. (a) The number of seats is the sum of the standard quotas.
Number of seats $= 41.2 + 31.9 + 24.8 + 22.6 + 16.5 = 137$

(b) The standard divisor is the total population divided by the number of available seats (found in part (a)).

$$\text{Standard divisor} = \frac{27,400,000}{137} = 200,000$$

(c) Population = standard quota × standard divisor, so
A: $41.2 \times 200,000 = 8,240,000$; B: $31.9 \times 200,000 = 6,380,000$; C: $24.8 \times 200,000 = 4,960,000$;
D: $22.6 \times 200,000 = 4,520,000$; E: $16.5 \times 200,000 = 3,300,000$

7. With 8.14% of the U.S. population, Texas should receive 8.14% of the number of seats available in the House of Representatives. So, the standard quota for Texas is $0.0814 \times 435 = 35.409 \approx 35.41$.

9. (a) Standard divisor $= \dfrac{100\%}{200} = 0.5\%$

(b) Alpha Kappa: $\dfrac{11.37\%}{0.5\%} = 22.74$; Beta Theta: $\dfrac{8.07\%}{0.5\%} = 16.14$; Chi Omega: $\dfrac{38.62\%}{0.5\%} = 77.24$;

Delta Gamma: $\dfrac{14.98\%}{0.5\%} = 29.96$; Epsilon Tau: $\dfrac{10.42\%}{0.5\%} = 20.84$; Phi Sigma: $\dfrac{16.54\%}{0.5\%} = 33.08$

4.2. Hamilton's Method

11. The standard quotas were calculated in Exercise 1(b) to be *A*: 66.2; *B*: 53.4; *C*: 26.6; *D*: 13.8. Lower quotas (the standard quotas rounded down) are thus *A*: 66; *B*: 53; *C*: 26; *D*: 13, and their sum is 158. So there are 160 - 158 = 2 seats remaining to allocate. These are given to *D* and *C*, since they have the largest fractional parts of the standard quota. The final Hamilton apportionment is *A*: 66; *B*: 53; *C*: 27; *D*: 14.

13. The standard quotas were calculated in Exercise 3(c) to be *A*: 43.558; *B*: 29.875; *C*: 19.702; *D*: 13.615; *E*: 9.865; *F*: 8.385. Lower quotas are thus *A*: 43; *B*: 29; *C*: 19; *D*: 13; *E*: 9; *F*: 8, and the sum is 121. So we have 125–121 = 4 buses remaining to allocate. These are given, in order, to *B, E, C* and *D*, since they have the largest fractional parts of the standard quota. The final Hamilton apportionment is *A*: 43; *B*: 30; *C*: 20; *D*: 14; *E*: 10; *F*: 8.

15. The standard quotas were given in Exercise 5 to be *A*: 41.2; *B*: 31.9; *C*: 24.8; *D*: 22.6; *E*: 16.5. Lower quotas (the standard quotas rounded down) are thus *A*: 41; *B*: 31; *C*: 24; *D*: 22; *E*: 16, and their sum is 134. So there are 137 - 134 = 3 seats remaining to allocate. These are given to *B, C* and *D*, since they have the largest fractional parts of the standard quota. The final Hamilton apportionment is *A*: 41; *B*: 32; *C*: 25; *D*: 23; *E*: 16.

17. The standard quotas were calculated in Exercise 9(b) to be *A*: 22.74; *B*: 16.14; *C*: 77.24; *D*: 29.96; *E*: 20.84; *F*: 33.08. Lower quotas are thus *A*: 22; *B*: 16; *C*: 77; *D*: 29; *E*: 20, *F*: 33, and the sum is 197. So we have 200–197 = 3 seats remaining to allocate. These are given to *D, E* and *A*, since they have the largest fractional parts of the standard quota. The final Hamilton apportionment is *A*: 23; *B*: 16; *C*: 77; *D*: 30; *E*: 21; *F*: 33.

19. (a) The standard divisor is $\dfrac{41 + 106 + 253}{24} = 16.667$. The standard quotas are then Dunes: 2.46; Smithville: 6.36; Johnstown: 15.18. Lower quotas are thus Dunes: 2; Smithville: 6; Johnstown: 15, and the sum is 23. So we have 24–23 = 1 social worker remaining to allocate. This is given to Dunes, since they have the largest fractional part of the standard quota. The final Hamilton apportionment is Dunes: 3; Smithville: 6; Johnstown: 15.

(b) The standard divisor is now $\dfrac{41 + 106 + 253}{25} = 16$. The standard quotas are now Dunes: 2.56; Smithville: 6.63; Johnstown: 15.81. Lower quotas are now Dunes: 2; Smithville: 6; Johnstown: 15, and the sum is 23. So we have 25–23 = 2 social workers remaining to allocate. These are given to Johnstown and Smithville, since they have the largest fractional part of the standard quota. The final Hamilton apportionment is Dunes: 2; Smithville: 7; Johnstown: 16.

(c) Dunes' apportionment went from 3 social workers to 2 even as the total number of social workers to be assigned increased from 24 to 25. This is an example of the Alabama paradox (with Dunes playing the role of Alabama).

4.3. Jefferson's Method

21. (a) The standard divisor is $\dfrac{4{,}500{,}000 + 4{,}900{,}000 + 3{,}900{,}000 + 6{,}700{,}000}{100} = 200{,}000$. To determine each county's standard quota, we divide that county's population by the standard divisor.
Arcadia: $\dfrac{4{,}500{,}000}{200{,}000} = 22.5$; Belarmine: $\dfrac{4{,}900{,}000}{200{,}000} = 24.5$;
Crowley: $\dfrac{3{,}900{,}000}{200{,}000} = 19.5$; Dandia: $\dfrac{6{,}700{,}000}{200{,}000} = 33.5$

(b) The lower quotas are Arcadia: 22; Belarmine: 24; Crowley: 19; Dandia: 33. The sum of these is 98.

(c) To determine each county's modified quota, we divide that county's population by the modified divisor.

Arcadia: $\dfrac{4,500,000}{197,000} = 22.84$; Belarmine: $\dfrac{4,900,000}{197,000} = 24.87$;

Crowley: $\dfrac{3,900,000}{197,000} = 19.80$; Dandia: $\dfrac{6,700,000}{197,000} = 34.01$

The modified quotas rounded down are now Arcadia: 22; Belarmine: 24; Crowley: 19; Dandia: 34. The sum of these is 99.

(d) Arcadia: $\dfrac{4,500,000}{195,000} = 23.08$; Belarmine: $\dfrac{4,900,000}{195,000} = 25.13$;

Crowley: $\dfrac{3,900,000}{195,000} = 20$; Dandia: $\dfrac{6,700,000}{195,000} = 34.36$

The modified quotas rounded down are now Arcadia: 23; Belarmine: 25; Crowley: 20; Dandia: 34. The sum of these is 102.

(e) Arcadia: $\dfrac{4,500,000}{195,800} = 22.98$; Belarmine: $\dfrac{4,900,000}{195,800} = 25.03$;

Crowley: $\dfrac{3,900,000}{195,800} = 19.92$; Dandia: $\dfrac{6,700,000}{195,800} = 34.22$

The modified quotas rounded down are now Arcadia: 22; Belarmine: 25; Crowley: 19; Dandia: 34. The sum of these is 100.

(f) Arcadia: $\dfrac{4,500,000}{196,000} = 22.96$; Belarmine: $\dfrac{4,900,000}{196,000} = 25$;

Crowley: $\dfrac{3,900,000}{196,000} = 19.90$; Dandia: $\dfrac{6,700,000}{196,000} = 34.18$

The modified quotas rounded down are now Arcadia: 22; Belarmine: 25; Crowley: 19; Dandia: 34. The sum of these is 100.

(g) We know that modified divisors $d = 195,800$ and $d = 196,000$ work from parts (e) and (f). Any other divisor in between (for example, $d = 195,900$) will also work.

23. Any modified divisor between approximately 49,285.72 and 49,402.98 can be used for this problem. Using $D = 49,300$ we obtain:

State	A	B	C	D
Modified Quota	67.14	54.16	26.98	13.996
Modified Lower Quota	67	54	26	13

The final apportionment using Jefferson's method is the modified lower quota value in the table.

25. Any modified divisor between approximately 1012 and 1024 can be used for this problem. Using $D = 1020$ we obtain:

Route	A	B	C	D	E	F
Modified Quota	44.412	30.461	20.088	13.882	10.059	8.549
Modified Lower Quota	44	30	20	13	10	8

The final apportionment using Jefferson's method is the modified lower quota value in the table.

27. Any modified divisor between approximately 196,200 and 196,500 can be used for this problem. Using $D = 196,500$ we obtain:

State	A	B	C	D	E
Modified Quota	41.93	32.47	25.24	23.00	16.79
Modified Lower Quota	41	32	25	23	16

The final apportionment using Jefferson's method is the modified lower quota value in the table.

29. Any modified divisor between approximately 0.4944% and 0.4951% can be used for this problem. Using $D = 0.495\%$ we obtain:

Planet	A	B	C	D	E	F
Modified Quota	22.9697	16.3030	78.0202	30.2626	21.0505	33.4141
Modified Lower Quota	22	16	78	30	21	33

The final apportionment using Jefferson's method is the modified lower quota value in the table.

4.4. Adams's and Webster's Method

31. Any modified divisor between approximately 205,500 and 205,800 can be used for this problem. Using $D = 205,600$ we obtain:

State	A	B	C	D	E
Modified Quota	40.08	31.03	24.12	21.98	16.05
Modified Upper Quota	41	32	25	22	17

The final apportionment using Adams's method is the modified upper quota value in the table.

33. Any modified divisor between approximately 0.5044% and 0.5081% can be used for this problem. Using $D = 0.505\%$ we obtain:

Planet	A	B	C	D	E	F
Modified Quota	22.5149	15.9802	76.4752	29.6634	20.6337	32.7525
Modified Upper Quota	23	16	77	30	21	33

The final apportionment using Adams's method is the modified upper quota value in the table.

35. Any modified divisor between 49,907 and 50,188 can be used for this problem. Using $D = 50,000$ we obtain:

State	A	B	C	D
Modified Quota	66.2	53.4	26.6	13.8
Rounded Modified Quota	66	53	27	14

The final apportionment using Webster's method is the rounded quota value in the table.

37. Any modified divisor between 200,100 and 200,800 can be used for this problem. Using $D = 200,500$ gives

State	A	B	C	D	E
Modified Quota	41.10	31.82	24.74	22.54	16.46
Rounded Modified Quota	41	32	25	23	16

The final apportionment using Webster's method is the rounded modified quota value in the table.

39. Any modified divisor between approximately 0.499% and 0.504% can be used for this problem. Using $D = 0.5\%$ we obtain:

State	A	B	C	D	E	F
Modified Quota	22.74	16.14	77.24	29.96	20.84	33.08
Rounded Modified Quota	23	16	77	30	21	33

The final apportionment using Webster's method is the rounded modified quota value in the table.

4.5. The Huntington-Hill Method

41. **(a)** 2. Since $1 \le 1.514 \le 2$, we need only compare 1.514 to the geometric mean $\sqrt{1 \times 2}$. Since $1.514 > \sqrt{2}$, we round up to 2. Note that $\sqrt{2} \approx 1.414213562$.

(b) 1. Since $1 \le 1.4 \le 2$, we need only compare 1.4 to the geometric mean $\sqrt{1 \times 2}$. Since $1.4 < \sqrt{2}$, we round down to 1.

(c) 1. As in (b), $1.41 < \sqrt{2}$ so we round down to 1.

(d) 2. $1.415 > \sqrt{2}$ so we round up to 2.

(e) 2. $1.449 > \sqrt{2}$ so we round up to 2.

43. 2. In 2010, the Huntington-Hill method was used in apportionment of the U.S. House of Representatives. Since $1 \le 1.485313 \le 2$, we need only compare 1.485313 to the geometric mean $\sqrt{1 \times 2} = \sqrt{2} \approx 1.414213562$. Since $1.485313 > \sqrt{2}$, we round up to 2. [See also Exercise 41.]

45. **(a)** The number of seats is the sum of the standard quotas.
Number of seats $= 3.52 + 10.48 + 1.41 + 12.51 + 12.08 = 40$.

(b) We start by attempting to round the standard quotas using the Huntington-Hill method. We calculate cutoffs and determine the rounded quotas sum to $4 + 10 + 1 + 13 + 12 = 40$. Therefore this is the final Huntington-Hill apportionment A: 4; B: 10; C: 1; D: 13; E: 12. [Note: We got somewhat lucky in not needing to modify the quotas.]

State	Standard Quota	Cutoff	Rounded Quota
A	3.52	3.464101615	4
B	10.48	10.48808848	10
C	1.41	1.414213562	1
D	12.51	12.489996	13
E	12.08	12.489996	12

47. (a) The number of seats is the sum of the standard quotas = $3.46 + 10.49 + 1.42 + 12.45 + 12.18 = 40$.

(b) We start by attempting to round the standard quotas using the Huntington-Hill method. We calculate cutoffs and determine the rounded quotas sum to $3 + 11 + 2 + 12 + 12 = 40$. Therefore this is the final Huntington-Hill apportionment *A*: 3; *B*: 11; *C*: 2; *D*: 12; *E*: 12. [Compare this to Exercise 45.]

State	Standard Quota	Cutoff	Rounded Quota
A	3.46	3.464101615	3
B	10.49	10.48808848	11
C	1.42	1.414213562	2
D	12.45	12.489996	12
E	12.18	12.489996	12

49. (a) A modified divisor of $d = 10000$ (the same as the standard divisor) works for this problem. The final apportionment under Webster's method is *A*: 34; *B*: 41; *C*: 22; *D*: 59; *E*: 15; *F*: 29.

State	Population	Modified Quota ($d = 10000$)	Rounded Quota
A	344,970	34.497	34
B	408,700	40.87	41
C	219,200	21.92	22
D	587,210	58.721	59
E	154,920	15.492	15
F	285,000	28.5	29

(b) We start by attempting to round the standard quotas using the Huntington-Hill method. We calculate cutoffs and determine the rounded quotas sum to $35 + 41 + 22 + 59 + 16 + 29 = 202$ (see table below). We need to modify the standard divisor upward.

State	Population	Standard Quota	Cutoff	Rounded Quota
A	344,970	34.497	34.4964	35
B	408,700	40.87	40.4969	41
C	219,200	21.92	21.4942	22
D	587,210	58.721	58.4979	59
E	154,920	15.492	15.4919	16
F	285,000	28.5	28.4956	29

We modify the divisor by choosing $d = 10001$. We calculate cutoffs and determine the rounded quotas sum to $35 + 41 + 22 + 59 + 15 + 29 = 200$. Therefore this is the final Huntington-Hill apportionment A: 34; B: 41; C: 22; D: 59; E: 15; F: 29.

State	Population	Modified Quota ($d = 10001$)	Cutoff	Rounded Quota
A	344,970	34.4936	34.4964	34
B	408,700	40.8659	40.4969	41
C	219,200	21.9178	21.4942	22
D	587,210	58.7151	58.4979	59
E	154,920	15.4905	15.4919	15
F	285,000	28.4972	28.4956	29

(c) The apportionments are the same under both methods.

4.6   The Quota Rule and Apportionment Paradoxes

51. The correct answer is (c). Under Hamilton's method there can be no violations of the quota rule. The only possible apportionment to state X is 35 or 36 (the value of the standard quota rounded down or up respectively). That is, the final apportionment using Hamilton's method must give X a (whole) number of seats less than one unit from its standard quota of 35.41.

53. The correct answer is (e). The other choices represent lower-quota violations and Jefferson's method can produce only upper-quota violations.

55. The correct answer is (e). Under Webster's method both lower-quota and upper-quota violations are possible.

57. The difference between 55 (the number of seats that California would receive under Jefferson's method) and 52.45 (California's standard quota) is greater than 1. This fact illustrates that Jefferson's method violates the quota rule (upper-quota violations are possible).

59. The Alabama paradox. The apportionment to Dunes went from 3 social workers to 2 for no other reason than the fact that the county hired an additional social worker (see answer to Exercise 19).

61. (a)

Child	Bob	Peter	Ron
Standard quota	0.594	2.673	7.733
Lower quota	0	2	7

Note: The standard divisor is $\dfrac{54 + 243 + 703}{11} = 90.91$.

The sum of lower quotas is 9, so there are 2 remaining pieces of candy to allocate. These are given to Ron and Peter, since they have the largest fractional parts of the standard quota. The final apportionment is: Bob: 0; Peter: 3; Ron: 8.

(b)

Child	Bob	Peter	Ron
Study time	56	255	789
Standard quota	.56	2.55	7.89
Lower quota	0	2	7

Note: The standard divisor is $\dfrac{56 + 255 + 789}{11} = 100$.

The sum of the lower quotas is 9, so there are 2 remaining pieces of candy to allocate. These are given to Ron and Bob, since they have the largest fractional parts of the standard quota. The final apportionment is: Bob: 1; Peter: 2; Ron: 8.

(c) Yes. For studying an extra 2 minutes (an increase of 3.70%), Bob gets a piece of candy. However, Peter, who studies an extra 12 minutes (an increase of 4.94%), has to give up a piece. This is an example of the population paradox.

63. (a) Standard divisor $= \dfrac{5,200,000 + 15,100,000 + 10,600,000}{50} = 618,000$.

(b) The standard quotas are Aila: 8.41; Balin: 24.43; Cona: 17.15. Lower quotas are Aila: 8; Balin: 24; Cona: 17 and the sum is 49. So we have 50–49 = 1 seat remaining to allocate. This is given to Balin since it has the largest fractional part of the standard quota. The final Hamilton apportionment is Aila: 8; Balin: 25; Cona: 17.

(c) The standard divisor is $\dfrac{5,200,000 + 15,100,000 + 10,600,000 + 9,500,000}{65} = 621,538.5$. The standard quotas are Aila: 8.37; Balin: 24.29; Cona: 17.05; Dent:15.28. Lower quotas are Aila: 8; Balin: 24; Cona: 17; Dent: 15 and the sum is 64. So we have 65–64 = 1 seat remaining to allocate. This is given to Aila since it has the largest fractional part of the standard quota. The final Hamilton apportionment is Aila: 9; Balin: 24; Cona: 17; Dent: 15.

(d) The new-states paradox. The addition of Dent (and the addition of the exact number of seats that Dent is entitled to) still impacts the other apportionments. Balin gives up one seat to Aila with the addition of a new state.

65. (a) Choose, for example, a modified divisor of $d = 990$ (the standard divisor is 1000. The rounded quotas sum to 88 + 4 + 5 + 3 = 100. Therefore Webster's apportionment is *A*: 88; *B*: 4; *C*: 5; *D*: 3.

State	Population	Modified Quota ($d = 990$)	Cutoff	Rounded Quota
A	86,915	87.7929	87.4986	88
B	4,325	4.3687	4.4721	4
C	5,400	5.4545	5.4772	5
D	3,360	3.3939	3.4641	3

(b) We start by attempting to round the standard quotas using the Huntington-Hill method. We calculate cutoffs and determine the rounded quotas sum to $87 + 4 + 5 + 3 = 99$ (see table below). We need to modify the standard divisor downward.

State	Population	Standard Quota	Cutoff	Rounded Quota
A	86,915	86.915	86.4986	87
B	4,325	4.325	4.4721	4
C	5,400	5.4	5.4772	5
D	3,360	3.36	3.4641	3

We modify the divisor by choosing $d = 990$. We calculate cutoffs and determine the rounded quotas sum to $88 + 4 + 5 + 3 = 100$. Therefore this is the final Huntington-Hill apportionment A: 88; B: 4; C: 5; D: 3.

State	Population	Modified Quota ($d = 990$)	Cutoff	Rounded Quota
A	86,915	87.7929	87.4986	88
B	4,325	4.3687	4.4721	4
C	5,400	5.4545	5.4772	5
D	3,360	3.3939	3.4641	3

(c) Both Webster's method and the Huntington-Hill method violate the quota rule. (An upper-quota violation occurs with A's apportionment – A's standard quota is only 86.915)

JOGGING

67. (a) State X has received at least one more seat than the standard quota suggested; an upper quota violation.

(b) State X has received at least one fewer seat than the standard quota suggested; a lower quota violation.

(c) The number of seats that state X has received is within 0.5 of its standard quota. Except for the case in which the standard quota has 0.5 as its decimal part, the number of seats that state X receives is the same as that found by using conventional rounding of the standard quota.

(d) The number of seats that state X has received satisfies the quota rule, but is not the result of conventional rounding of the standard quota.

69. If the modified divisor is d, then state A has modified quota $\dfrac{p_A}{d} \geq 2.5$. Also, state B has modified quota $\dfrac{p_B}{d} < 0.5$. So, $p_A \geq 2.5d > 5p_B$. That is, more than $\dfrac{5}{6}$ of the population lives in state A.

71. (a) In Jefferson's method the modified quotas are larger than the standard quotas and so rounding downward will give each state at least the integer part of the standard quota for that state.

(b) In Adam's method the modified quotas are smaller than the standard quota and so rounding upward will give each state at most one more than the integer part of the standard quota for that state.

(c) If there are only two states, an upper quota violation for one state results in a lower quota violation for the other state. Neither Jefferson's nor Adams' method can have both upper and lower quota violations. So, when there are only two states neither method can violate the quota rule.

73. (a) *A*: 33; *B*: 138; *C*: 4; *D*: 41; *E*: 14; *F*: 20

(b) State *D* is apportioned 41 seats under Lowndes' method while all of the methods discussed in the chapter apportion 42 seats to state *D*. States *C* and *E* do better under Lowndes' method than under Hamilton's method.

RUNNING

75. (a) We use a standard divisor of $SD = \dfrac{200,000}{22} \approx 9090.91$. The standard quotas are given below.

State	*A*	*B*	*C*	*D*
Population	18,000	18,179	40,950	122,871
Standard Quota	1.98	1.9997	4.5045	13.5158

Each state would then receive its lower quota: *A*: 1; *B*: 1; *C*: 4; *D*: 13. This leads to 3 surplus seats to be divided among the four states. To apportion these surplus seats, we use Jefferson's method. A modified divisor of *d* = 40,955 works to apportion these three surplus seats.

State	*A*	*B*	*C*	*D*
Population	18,000	18,179	40,950	122,871
Modified Quota (*d* = 40,955)	0.4395	0.4439	0.9999	3.0001
Lower Quota	0	0	0	3

The final apportionment would have state *D* receiving the 3 surplus seats: *A*: 1; *B*: 1; *C*: 4; *D*: 16.

(b) The Hamilton-Jefferson hybrid method can produce apportionments different from Hamilton's method because, as is clear from (a), the quota rule can be violated in the hybrid method.
For the example in (a), the apportionment due to Jefferson's method is: *A*: 2; *B*: 2; *C*: 4; *D*: 14. This example shows that the Hamilton-Jefferson hybrid method can produce apportionments different from Jefferson's method. The bias towards large states present in Hamilton's method and in Jefferson's method are both being exploited to make this method even more biased towards large states.

(c) The hybrid method can violate the quota rule since more than one surplus seat can be handed out in the second step of the method (the Jefferson step).

77. (a) Under Adams' method, an increase in the total number of seats to be apportioned results in a decrease in the modified divisor which will result in some of the states gaining seats, but none will lose seats (since all modified quotas will increase).

(b) Under Adams' method, the addition of a new state with its fair share of seats may require a new modified divisor. If the modified divisor is increased, some original states may lose seats, but none will gain seats; if the modified divisor is decreased, some original states may gain seats, but none will lose seats. That is, no state will gain a seat at the expense of another.

Chapter 5

WALKING

5.2   An Introduction to Graphs

1. **(a)** $\{A, B, C, X, Y, Z\}$
    The basic elements of a graph are dots and lines (vertices and edges in mathematical language). The vertex set is a listing of the labels given to each vertex (dot).

    (b) $\{AX, AY, AZ, BB, BC, BX, CX, XY\}$
    Remember, each edge can be described by listing the pair of vertices that are connected by that edge.

    (c) $\deg(A) = 3$, $\deg(B) = 4$, $\deg(C) = 2$, $\deg(X) = 4$, $\deg(Y) = 2$, $\deg(Z) = 1$
    The degree of a vertex is the number of edges meeting at that vertex.

    (d) One possible answer:

    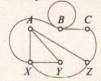

3. **(a)** $\{A, B, C, D, X, Y, Z\}$

    (b) $\{AX, AX, AY, BX, BY, DZ, XY\}$
    Note: AX is listed twice.

    (c) $\deg(A) = 3$, $\deg(B) = 2$, $\deg(C) = 0$, $\deg(D) = 1$, $\deg(X) = 4$, $\deg(Y) = 3$, $\deg(Z) = 1$

    (d) There are 3 components in this graph (one component contains the vertices A, B, X, and Y and each edge connecting them; one component contains the vertex C; one component contains vertices D and Z and the edge that connects them).

5. One possible picture of the graph:

    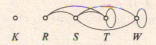

    A second possible picture:

7. **(a)** A, B, D, E
    Remember that two vertices are adjacent in a graph if they are joined by an edge. So we may simply consider the edges that contain a D (listed either first or second).

    (b) AD, BC, DD, DE
    Two edges are adjacent if they share a common vertex (in this case, either vertex B or vertex D).

    (c) 5; D appears 5 times in the edge list.

    (d) 12; one way to easily find this is to multiply the number of edges by two (since each edge contributes two degrees to the total).

9. **(a)** Arranging the vertices in a circular pattern and connecting them with edges that run "around the circle" will accomplish this task.

    (b) One method is to construct two copies of the graph in part (a) – each on 4 vertices.

    (c) One method is to arrange the vertices in two equally long rows and connect each vertex in the first row with exactly one vertex in the second.

11. (a) *C, B, A, H, F*
Other answers such as *C, B, A, H, G, F* are also possible.

    (b) *C, B, D, A, H, F*
Again, more than one answer is possible.

    (c) *C, B, A, H, F*

    (d) *C, D, B, A, H, G, G, F*

    (e) 4 (It can be helpful to list them from shortest to longest as *C, B, A*; *C, D, A*; *C, B, D, A*; *C, D, B, A*)

    (f) 3 (*H, F*; *H, G, F*; *H, G, G, F*)

    (g) 12 (Any one of the 4 paths in (e) followed by edge *AH* followed by any one of the 3 paths in (f).)

13. (a) *G,G*. A loop is a circuit of length 1.

    (b) There are none. Such a circuit would consist of two vertices and two *different* edges connecting the vertices. [This explains why *G,G,G* is not a circuit of length 2.]

    (c) *A,B,D,A; B,C,D,B; F,G,H,F*
For starters, look for triangles consisting of three vertices and three different edges in the graph.

    (d) *A,B,C,D,A; F,G,G,H,F*
For starters, look for rectangles consisting of four vertices and four different edges such as *A,B,C,D,A* in the graph. Note that other possibilities (such as the triangle with a loop *F,G,G,H,F*) also exist.

15. (a) *AH, EF*; If either of these edges were removed from the graph, the graph would be disconnected.

    (b) There are none. This graph is just one circuit connecting five vertices. Circuits do not have bridges.

    (c) *AB, BC, BE, CD*; That is, every edge is a bridge.

17. (a) *CI* and *HJ*. Start by noting the large clique formed by *A*, *B*, *C*, *D*, and *E*. A second clique is formed by *F*, *G*, and *I* and a third clique is formed by *G*, *H*, and *I*. Since cliques do not have bridges, only edges *CI* and *HJ* are possible bridges.

    (b) There would be 3 components in such a graph. One component would contain the clique formed by *A*, *B*, *C*, *D*, and *E*; one component would contain vertices *F*, *G*, *H*, and *I* and the edges that connect them; one component would contain vertex *J*.

    (c) The shortest path is *C, I, H, J* of length 3.

    (d) A longest path is *I, G, F, I, H, J*. It has length 5. Another longest path is *I, F, G, I, H, J*. A longer path is not possible since it cannot traverse through any of the vertices *A*, *B*, *C*, *D*, and *E* (if it did, it would need to traverse edge *CI* twice which isn't allowed).

19. Such a graph will have five vertices (representing North Kingsburg, South Kingsburg, and islands *A*, *B*, and *C*) and seven edges (representing the seven bridges).

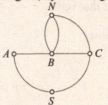

21.

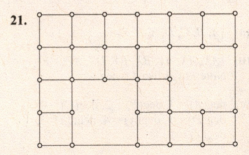

23. Let the vertices represent the teams *A, B, C, D, E,* and *F*. The edges will correspond to the tournament pairings. Putting the vertices at the corners of a regular hexagon will make drawing and interpreting the graph a bit easier.

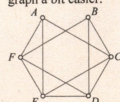

25. First, draw the vertices (these represent the people). Then, draw the edges (when present) which represent Facebook friendships.

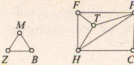

27. First, draw the vertices (these represent the intersections of two streets). Then, draw the edges (representing streets).

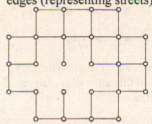

5.3 Euler's Theorems and Fleury's Algorithm

29. (a) (A); The graph has an Euler circuit because it is connected and all vertices have even degree (Euler's Circuit Theorem).

(b) (C); The graph has neither an Euler circuit nor an Euler path because there are four vertices of odd degree (Euler's Path Theorem).

(c) (A); Exercise 9(a) gave a representation of this graph. We can then apply Euler's Circuit Theorem.

(d) (D); Exercises 9(a) and 9(b) gave two possible representations of such a graph. In Exercise 9(a), the graph is connected so that Euler's Circuit Theorem guarantees that it has an Euler Circuit.

In Exercise 9(b), the graph is not connected and does not have an Euler Circuit.

(e) (F); The sum of the degrees of all the vertices of this graph is $3 \times 2 + 5 \times 3 = 21$ which is odd. Euler's Sum of Degrees Theorem tells us that this cannot happen in any graph.

31. (a) (C); The graph has neither an Euler circuit nor an Euler path because there are more than two (exactly 4 in fact) vertices of odd degree (Euler's Path Theorem).

(b) (A); The graph has an Euler circuit because all vertices have even degree (Euler's Circuit Theorem).

(c) (F); The sum of the degrees of all the vertices of this graph is $11 \times 3 = 33$ which is odd. Euler's Sum of Degrees Theorem tells us that this cannot happen in any graph.

(d) (C); The graph has neither an Euler circuit nor an Euler path since it is connected and exactly 12 vertices are odd (Euler's Path Theorem).

33. (a) (B); The graph has an Euler path since it is connected and exactly two vertices are odd (Euler's Path Theorem).

(b) (A); The graph has an Euler circuit because all vertices have even degree (Euler's Circuit Theorem).

(c) (C); While all the vertices have even degree, the vertex of degree zero tells us that the graph is disconnected. So Euler's Path Theorem tells us that the graph has neither an Euler circuit nor an Euler path.

35.

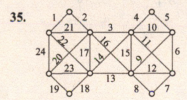

37. The starting and ending vertices of the Euler path (both having odd degree of course!) are shown in black. Naturally, other answers are possible.

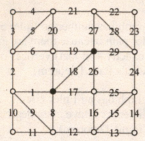

39. There is more than one answer.

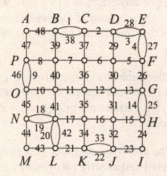

41. (a) *PD* or *PB*; Edges *PA* and *PE* have already been traveled. Neither *PD* nor *PB* is a bridge of the yet-to-be-traveled part.

(b) *BF*; *BF* is a bridge of the yet-to-be-traveled part. If *BF* were removed, you would be left with the following components.

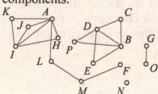

5.4   Eulerizing and Semi-Eulerizing Graphs

43. Adding as few duplicate edges as possible in eliminating the odd vertices gives an optimal eulerization.

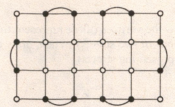

45. Since this graph has vertical symmetry, a mirror image of this answer is also an optimal eulerization.

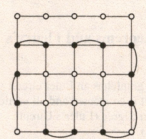

47. In an optimal semi-eulerization two vertices will remain odd. Again, since this graph has vertical symmetry, a mirror image of this answer is also an optimal semi-eulerization. Other answers are possible.

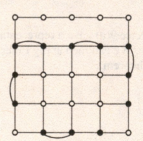

49. In an optimal semi-eulerization, the two vertices *A* and *D* will remain odd. The vertex between *A* and *D* then requires some work to become even (i.e. it is not next to an odd vertex). Other answers are possible.

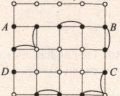

51. In an optimal semi-eulerization, the two vertices B and C will remain odd. The vertex between B and C then requires at least two additional edges to become even (i.e. it is not right next to an odd vertex).

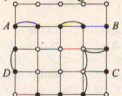

Other answers are possible. One of these is shown below.

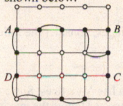

53. Adding as few duplicate edges as possible in eliminating the odd vertices gives an optimal eulerization. Other answers are possible.

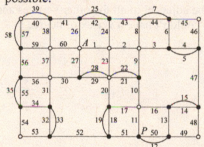

55. In an optimal semi-eulerization, the two vertices A and B will remain odd. Other answers are possible.

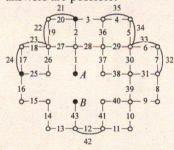

JOGGING

57. 5; There are 12 odd vertices in the figure. Two vertices can be used as the starting and ending vertices. The remaining 10 vertices can be paired so that each pair forces one lifting of the pencil.

59. None. If a vertex had degree 1, then the single edge incident to that vertex would be a bridge.

61. Each component of the graph is a graph in its own right and so, according to Euler's Sum of Degrees Theorem, the number of vertices of odd degree (in each component) must be even. Therefore the 2 vertices of odd degree must be in the same component.

63. (a) *NK, C, B, NK, B, A, SK, C*

   (b) *NK, C, B, NK, B, A, SK, C, SK*

   (c) *NK, C, B, NK, B, A, SK, C, B*

   (d) *NK, C, B, NK, B, A, SK, C, B, A*

65. (a) There are many ways to implement this algorithm. The following figures illustrate one possibility.

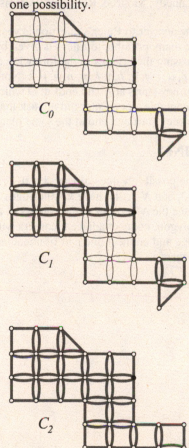

(b) Step 1: Find a path C_0 between the two vertices v and w of odd degree.

Step 2: Find a circuit C_0^* that kisses C_0 at any vertex that is on C_0. The path and the circuit can be combined into a larger path C_1 between the two odd vertices. If there are no kissing circuits to the larger path, we are done.

Step 3: Repeat Step 2 until an Euler path is found.

67. Since vertices R, B, D, and L are all odd, two bridges must be crossed twice. For example, crossing the Adams Bridge twice and the Lincoln Bridge twice produces an optimal route: $D, L, R, L, C, A, R, B, R, B, D, C, L, D$. Listing the bridges in the order they are crossed: Adams, Washington, Jefferson, Grant, Wilson, Truman, Lincoln, Lincoln, Hoover, Kennedy, Monroe, Roosevelt, Adams.

69. The answer to Euler's question is yes. One of the many possible journeys is given by crossing the bridges in the following order: a, b, c, d, e, f, g, h, i, l, m, n, o, p, k. Note that this journey starts at E and ends at D which is acceptable since Euler did not ask that the journey start and end at the same place.

RUNNING

71. The possible values of k are $k = 0, 1, 2, 3, \ldots$, N-3, and N-1. To have $k = 0$ bridges in G, place the N vertices at the corners of a polygon, connect adjacent corners with an edge and connect one pair of nonadjacent corners.

To have $k = 1$ bridge, attach a single vertex to a graph formed by putting vertices at the corners of a regular N-1 polygon.

To have $k = 2$ bridges, consider the following graph formed by attaching two vertices to a graph formed by putting vertices at the corners of a regular N-2 polygon. Adding more "single" vertices to the left also explains how to get $k = 3, 4, \ldots, N$-3 bridges.

To get $k = N$-1 bridges, consider the following example:

73. Suppose the graph has N vertices. Since there are no multiple edges or loops, the maximum degree a vertex can have is $N - 1$. If the degrees of the N vertices are all different, they must be $0, 1, 2, \ldots, N - 1$, but this is impossible because the vertex of degree $N - 1$ would have to be adjacent to all the other vertices and then we couldn't have a vertex of degree 0.

Chapter 6

WALKING

6.2 Hamilton Paths and Circuits

1. **(a)** Remember that a Hamilton circuit is a circuit that includes each vertex of the graph once and only once. Remember too that the circuit must return to the starting vertex.
   1. A, B, D, C, E, F, G, A;
   2. A, D, C, E, B, G, F, A;
   3. A, D, B, E, C, F, G, A

   (b) Remember that a Hamilton path in a graph is a path that includes each vertex of the graph once and only once. Unlike a circuit, it need not end where it started.
   A, G, F, E, C, D, B

   (c) D, A, G, B, C, E, F

3. Starting at A, suppose that you travel to B. Then, there are two choices for which vertex to travel to next: C and E. If one starts out A, B, C, then the next vertex visited must be D (do you see why? Otherwise you would eventually pass through C twice which is not allowed). If one starts out A, B, E, then the next vertex visited must also be D. After D, there is only one way to return to A to form a Hamilton circuit. So the Hamilton circuits are A, B, C, D, E, F, G, A and A, B, E, D, C, F, G, A and their mirror images. [Note: The mirror images are A, G, F, E, D, C, B, A and A, G, F, C, D, E, B, A.]

5. **(a)** A, F, B, C, G, D, E;
   Note that edges $AF, FB, CG,$ and GD (or their mirror images) must be part of a Hamilton path.

   (b) A, F, B, C, G, D, E, A

   (c) One approach is to "spiral" (in a counterclockwise fashion) around from E to A. One such path is E, D, G, C, B, F, A.

   (d) The Hamilton path found in part (c) can be easily modified to do the job. One Hamilton circuit is E, D, G, C, B, F, A, E.

7. **(a)** 8; Knowing that in a Hamilton circuit, each vertex is listed once and only once, we simply count the number of unique vertices listed in the given Hamilton circuit.

(b) $A, H, C, B, F, D, G, E, A$

(c) We can write a Hamilton path starting at A using the circuit found in (b) that does not return back to A. That is, A, H, C, B, F, D, G, E. Of course, the path A, E, G, D, F, B, C, H found using the mirror image from (b) will also work.

9. **(a)** Breaking between C and B gives B, A, D, E, C.
   Breaking between E and C gives C, B, A, D, E.
   Breaking between D and E gives E, C, B, A, D.
   Breaking between A and D gives D, E, C, B, A.
   Breaking between B and A gives A, D, E, C, B.

   (b) As described in Example 6.6, A, B, E, D, C is one such path. The mirror image C, D, E, B, A is another.

   Another path that does not come from a "broken" Hamilton circuit is D, A, E, C, B since we cannot get back to D directly from B. Naturally the mirror image B, C, E, A, D is another path.

   Rather than these "Z" patterns as in the previous paths, we can construct "backwards Z" patterns to find four additional paths: B, A, E, C, D and its mirror image D, C, E, A, B and A, D, E, B, C and its mirror image C, B, E, D, A.

11. **(a)** Suppose that starting at A we move toward F. The next vertex visited must then be E. After that, we have two choices (D or B). But once D or B is chosen, the rest of the circuit is forced. So we get A, F, E, D, C, B, A or A, F, E, B, C, D, A (or the reversals).

(b) The idea here is to start at C and end at F so that there is no single edge that would connect back to C. From C, we can travel to B and then choose to go to A or E. To end at a point other than D, the path is unique.

C, B, A, D, E, F;
C, B, E, D, A, F;

From C we could also travel in the other direction (to D rather than B first) and get paths

C, D, E, B, A, F;
C, D, A, B, E, F.

13. (a) When constructing such a path, be sure to travel to all vertices A, B, C, D, and E before crossing edge CI. Also note that from A only vertices B or D (and not E) may be next. One such path is A, B, E, D, C, I, H, G, K, J, F.

(b) In this case, be sure to travel to all of the vertices K, J, F, G, H, and I before crossing edge IC. Starting at K, there is only one way to travel to I. This gives one path $K, J, F, G, H, I\,C, B, A, D, E$. Another possible path is $K, J, F, G, H, I\,C$, D, A, B, E.

(c) CI is a bridge of the graph connecting a "left half" and a "right half." If you start at C and go left, there is no way to get to the right half of the graph without going through C again. On the other hand, if you start at C and cross over to the right half first (i.e. you move to vertex I), there is no way to get back to the left half without going through C again.

(d) No matter where you start, you would have to cross the bridge CI twice to visit every vertex and get back to where you started. This is a problem since you would then hit either vertex C or vertex I twice within the circuit.

15. There is no Hamilton circuit since two vertices have degree 1. There is no Hamilton path since any such path must contain edges AB, BE, and BC, which would force vertex B to be visited more than once.

17. (a) 6

(b) B, D, A, E, C, B
Weight $= 6 + 1 + 9 + 4 + 7 = 27$

(c) The mirror image B, C, E, A, D, B
Weight $= 7 + 4 + 9 + 1 + 6 = 27$

19. (a) A, D, F, E, B, C
Weight $= 2 + 7 + 5 + 4 + 11 = 29$

(b) A, B, E, D, F, C
Weight $= 10 + 4 + 3 + 7 + 6 = 30$

(c) There are only two Hamilton paths that start at A and end at C. The path $A, D, F,$ E, B, C found in part (a) is the optimal such path. It has weight 29.

21. (a) 20! circuits $\times \dfrac{1 \text{ sec}}{1,000,000,000 \text{ circuits}}$

$\times \dfrac{1 \text{ min}}{60 \text{ sec}} \times \dfrac{1 \text{ hr}}{60 \text{ min}} \times \dfrac{1 \text{ day}}{24 \text{ hr}} \times \dfrac{1 \text{ year}}{365 \text{ days}}$

≈ 77 years

(b) 21! circuits $\times \dfrac{1 \text{ sec}}{1,000,000,000 \text{ circuits}}$

$\times \dfrac{1 \text{ min}}{60 \text{ sec}} \times \dfrac{1 \text{ hr}}{60 \text{ min}} \times \dfrac{1 \text{ day}}{24 \text{ hr}} \times \dfrac{1 \text{ year}}{365 \text{ days}}$

≈ 1622 years

23. (a) $\dfrac{20 \times 19}{2} = 190$

(b) K_{21} has 20 more edges than K_{20}.

$\dfrac{21 \times 20}{2} = 210$

(c) If one vertex is added to K_{50} (making for 51 vertices), the new complete graph K_{51} has 50 additional edges (one to each old vertex). In short, K_{51} has 50 more edges than K_{50}. That is, $y - x = 50$.

25. (a) K_N has $(N-1)!$ Hamilton circuits and $120 = 5!$. So $N = 6$.

(b) K_N has $\dfrac{(N-1)N}{2}$ edges and $45 = \dfrac{9 \times 10}{2}$, so $N = 10$.

(c) $20,100 = \dfrac{200 \times 201}{2}$, so $N = 201$.

6.3 The Brute Force Algorithm

27.

Tour	Weight	Mirror-image Circuit
A, B, C, D, A	$48 + 32 + 18 + 22 = 120$	A, D, C, B, A
A, B, D, C, A	$48 + 20 + 18 + 28 = 114$	A, C, D, B, A
A, C, B, D, A	$28 + 32 + 20 + 22 = 102$	A, D, B, C, A

An optimal tour would be A, C, B, D, A (or its reversal) with a cost of 102.

29.

Tour	Weight	Mirror-image Circuit
A, B, C, D, E, A	$35 + 9 + 25 + 32 + 18 = 119$	A, E, D, C, B, A
A, B, C, E, D, A	$35 + 9 + 23 + 32 + 22 = 121$	A, D, E, C, B, A
A, B, D, C, E, A	$35 + 21 + 25 + 23 + 18 = 122$	A, E, C, D, B, A
A, B, D, E, C, A	$35 + 21 + 32 + 23 + 40 = 151$	A, C, E, D, B, A
A, B, E, C, D, A	$35 + 18 + 23 + 25 + 22 = 123$	A, D, C, E, B, A
A, B, E, D, C, A	$35 + 18 + 32 + 25 + 40 = 150$	A, C, D, E, B, A
A, C, B, D, E, A	$40 + 9 + 21 + 32 + 18 = 120$	A, E, D, B, C, A
A, C, B, E, D, A	$40 + 9 + 18 + 32 + 22 = 121$	A, D, E, B, C, A
A, C, D, B, E, A	$40 + 25 + 21 + 18 + 18 = 122$	A, E, B, D, C, A
A, C, E, B, D, A	$40 + 23 + 18 + 21 + 22 = 124$	A, D, B, E, C, A
A, D, B, C, E, A	$22 + 21 + 9 + 23 + 18 = 93$	A, E, C, B, D, A
A, E, B, C, D, A	$18 + 18 + 9 + 25 + 22 = 92$	A, D, C, B, E, A

An optimal tour would be A, E, B, C, D, A (or its reversal) with time of 92 hours.

31.

Tour	Weight	Mirror-image Circuit
B, C, D, A, B	$17 + 11 + 14 + 12 = 54$	B, A, D, C, B
B, D, C, A, B	$15 + 11 + 6 + 12 = 44$	B, A, C, D, B
B, D, A, C, B	$15 + 14 + 6 + 17 = 52$	B, C, A, D, B

An optimal tour would be B, D, C, A, B (or its reversal) with a time of 44 hours.

6.4   The Nearest-Neighbor and Repetitive Nearest-Neighbor Algorithms

33. (a) Starting at B, the nearest neighbor is vertex C with cost of $121 (compared to costs of $185 to vertex A, $150 to vertex D, and $200 to vertex E). Once at C, we check the cost of flying to vertices A, D, and E. Of these, flying from C to A is the cheapest at $119. From A, we check the cost of flying to D and E. Since E is cheapest, we fly from A to E ($133) and then from E to D ($199). Finally, we complete the circuit and fly back to vertex B at a cost of $150. Putting this all together gives the following circuit and cost:
B, C, A, E, D, B; Cost = $121 + $119 + $133 + $199 + $150 = $722

(b) C, A, E, D, B, C; Cost = $119 + $133 + $199 + $150 + $121 = $722

(c) D, B, C, A, E, D; Cost = $150 + $121 + $119 + $133 + $199 = $722

(d) E, C, A, D, B, E; Cost = $120 + $119 + $152 + $150 + $200 = $741

35. (a) Starting at A, the nearest neighbor is vertex D with distance of 185 miles (compared to distances of 500 miles to vertex B, 200 miles to vertex C, and 205 miles to vertex E). Once at D, we check the distances of driving to vertices B, C, and E. Of these, driving from D to E is the shortest distance at 302 miles. From E, we check the distance to B and C. Since C is the shortest time, we drive from E to C (165 miles) and then from C to B (305 miles). Finally, we complete the circuit and drive back to vertex A at a distance of 500 miles. Putting this all together gives the following circuit and cost:
A, D, E, C, B, A; Cost of bus tour = $8 \times (185 + 302 + 165 + 305 + 500) = $11,656

(b) A, D, B, C, E, A
Cost of bus tour = $8 \times (185 + 360 + 305 + 165 + 205) = $9,760

37. (a) Starting at A, the nearest neighbor is vertex E with time of 18 hours (compared to times of 35 hours to vertex B, 40 hours to vertex C, and 22 hours to vertex D). Once at E, we check the travel time of driving to vertices B, C, and D. Of these, driving from E to B is the shortest time at 18 hours. From B, we check the travel time to C and D. Since C is the shortest time, we drive from B to C (9 hours) and then from C to D (25 hours). Finally, we complete the circuit and drive back to vertex A at a travel time of 22 hours. Putting this all together gives the following circuit and cost:
A, E, B, C, D, A; Total travel time = $18 + 18 + 9 + 25 + 22 = 92$ hours

(b) A, D, B, C, E, A; Total travel time = 93 hours

39. (a) The smallest number in the Atlanta column is for Columbus; the smallest number other than Atlanta in the Columbus column is for Kansas City; the smallest number in the Kansas City column other than Atlanta and Columbus is Tulsa, etc.. So the nearest-neighbor tour for Darren is Atlanta, Columbus, Kansas City, Tulsa, Minneapolis, Pierre, Atlanta. The cost of this trip = $0.75 \times (533 + 656 + 248 + 695 + 394 + 1361) = $2915.25.

(b) The nearest-neighbor circuit with Kansas City as the starting vertex is Kansas City, Tulsa, Minneapolis, Pierre, Columbus, Atlanta, Kansas City. Written with starting city Atlanta, the circuit is Atlanta, Kansas City, Tulsa, Minneapolis, Pierre, Columbus, Atlanta. The cost of this trip = $0.75 \times (798 + 248 + 695 + 394 + 1071 + 533) = \$2804.25$

41.

Starting Vertex	Tour	Length of Tour
A	A, E, B, C, D, A	$18 + 18 + 9 + 25 + 22 = 92$
B	B, C, E, A, D, B	$9 + 23 + 18 + 22 + 21 = 93$
C	C, B, E, A, D, C	$9 + 18 + 18 + 22 + 25 = 92$
D	D, A, E, B, C, D	$22 + 18 + 18 + 9 + 25 = 92$
E	E, A, D, B, C, E	$18 + 22 + 21 + 9 + 23 = 93$
	E, B, C, D, A, E	$18 + 9 + 25 + 22 + 18 = 92$

Starting with vertex A, the shortest circuit is A, E, B, C, D, A (or its reversal) with cost of 92 hours.

43.

Starting Vertex	Tour	Length of Tour (Miles)
A	A, C, K, T, M, P, A	$533 + 656 + 248 + 695 + 394 + 1361 = 3887$
C	C, A, T, K, M, P, C	$533 + 772 + 248 + 447 + 394 + 1071 = 3465$
K	K, T, M, P, C, A, K	$248 + 695 + 394 + 1071 + 533 + 798 = 3739$
M	M, P, K, T, A, C, M	$394 + 592 + 248 + 772 + 533 + 713 = 3252$
P	P, M, K, T, A, C, P	$394 + 447 + 248 + 772 + 533 + 1071 = 3465$
T	T, K, M, P, C, A, T	$248 + 447 + 394 + 1071 + 533 + 772 = 3465$

The shortest circuit is M, P, K, T, A, C, M. Written starting from Atlanta, this is Atlanta, Columbus, Minneapolis, Pierre, Kansas City, Tulsa, Atlanta and has a length of 3252 miles. The total cost of this tour is $0.75 \times 3252 = \$2439$.

45. The relative error of the nearest-neighbor tour is $\varepsilon = (W - Opt)/Opt$ where W is the weight of the nearest-neighbor tour and Opt is the weight of the optimal tour. In this case, we see that this error is

$\varepsilon = (W - Opt)/Opt = (\$13,500 - \$12,000)/\$12,000 = 12.5\%.$

6.5 Cheapest-Link Algorithm

47. The cheapest-link tour is created by adding edges BC, AE, BE, AD and CD in that order.

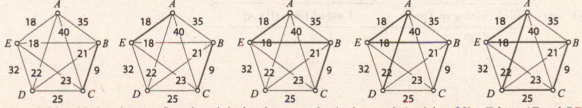

Edge BC is added to the tour first since it is the cheapest edge in the graph (weight of 9). Edges AE and BE are the next cheapest edges (weights of 18) and are added next. Edge BD is the next cheapest edge but it is not added to the tour since then three edges would meet at vertex B. Edge AD is added fourth with weight 22. Edge CE is next cheapest but it is not added to the tour since then three edges would meet at vertex E. Edge CD is added at this point and completes the tour. The shortest tour using this algorithm is A, D, C, B, E, A (or its reversal). The cost of this tour is $22 + 25 + 9 + 18 + 18 = 92$ hours.

49. The first three steps in the cheapest-link algorithm are illustrated in the following figures.

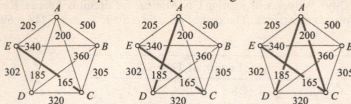

After the third step, the next cheapest edge is *AE* (weight 205). But that edge would make three edges come together at vertex *A* so we skip it and move to the next cheapest edge which is *DE* (weight 302). Since this edge makes a circuit, we skip this edge and try the next cheapest edge. The next cheapest edge is *BC* (weight 305) but that makes three edges come together at vertex *C* so we skip this edge and choose the next cheapest edge which is *CD* (weight 320). That edge makes a circuit and also makes three edges come together at vertex *C*. So, we skip it and choose the next cheapest edge which is *BE* (weight 340). After *BE* is added, only one edge, *BD*, will complete the tour.

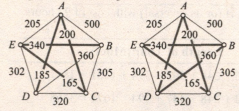

The shortest tour found using the cheapest-link algorithm is *A, D, B, E, C, A*. The cost of this bus tour is $8 \times (165 + 185 + 200 + 340 + 360) = \$10,000$.

51.

Link Added to Tour	Cost of Link (Miles)
Kansas City - Tulsa	248
Pierre - Minneapolis	394
Minneapolis - Kansas City	447
Atlanta - Columbus	533
Atlanta - Tulsa	772
Columbus - Pierre	1071

Darren's cheapest-link tour is Atlanta, Columbus, Pierre, Minneapolis, Kansas City, Tulsa, Atlanta. His total mileage is $248 + 394 + 447 + 533 + 772 + 1071 = 3465$ miles. His total cost is $\$0.75 \times 3465 = \$2598.75$.

53. (a)

Link Added to Tour	Cost of Link (Days)
A - E	9
A - F	10
C - F	11
D - E	12
B - D	13
B - C	20

The rover's cheapest-link tour is *A, E, D, B, C, F, A*. The length is $9 + 12 + 13 + 20 + 11 + 10 = 75$ days.

JOGGING

55. **(a)** $7! = 5040$

(b) $7! = 5040$; this is the same as in part (a). The remaining seven letters can be rearranged in any sequence.

(c) $6! = 720$; the remaining six letters can be arranged in any sequence.

57. We solve $(\$1614 - Opt)/Opt = 7.6\%$ for *Opt* where *Opt* is the cost of the optimal solution. But this gives $\$1614 - Opt = 0.076 \cdot Opt$ or $\$1614 = 0.076 \cdot Opt + Opt$. So,

$$Opt = \frac{\$1614}{1.076} = \$1500.$$

59. **(a)**

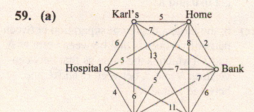

(b) Just "eyeballing it" will give the optimal tour in this case. It is: Home, Bank, Post Office, Deli, Hospital, Karl's , Home. The total length of the tour is $2 + 6 + 7 + 4 + 6 + 5 = 30$ miles.

61. The graph describing the friendships among the guests does not have a Hamilton circuit. Thus it is impossible to seat everyone around the table with friends on both sides.

63. Suppose the cheapest edge in a graph is the edge joining vertices *X* and *Y*. Using the nearest neighbor algorithm we will eventually visit one of these vertices—suppose the first one of these vertices we visit is *X*. Then, since edge *XY* is the cheapest edge in the graph and since we have not yet visited vertex *Y*, the nearest neighbor algorithm will take us to *Y*.

RUNNING

65. **(a)** Julie should fly to Detroit. The optimal route will then be to drive $68 + 56 + 68 + 233 + 78 + 164 + 67 + 55 = 789$ miles via Detroit-Flint-Lansing-Grand Rapids-Cheboygan-Sault Ste. Marie-Marquette-Escanaba-Menominee for a total cost of $789 \times \$0.39 + \$3.00 = \$310.71$. Since Julie can drive from Menominee back to Detroit (via Sault Ste. Marie and Cheboygan) in a matter of $227 + 78 + 280 = 585$ miles at a cost of $585 \times (\$0.39) + \$3.00 = \$231.15$, she should do so and drop the rental car back in Detroit (assuming she need not pay extra for gas!). The total cost of her trip would then be $\$310.71 + \$231.15 = \$541.86$.

(b) The optimal route would be for Julie to fly to Detroit and drive $68 + 56 + 68 + 233 + 78 + 164 + 67 + 55 = 789$ miles along the Hamilton path Detroit-Flint-Lansing-Grand Rapids-Cheboygan-Sault Ste. Marie-Marquette-Escanaba-Menominee for a total cost of $789 \times \$0.49 + \$3.00 = \$389.61$.

67. **(a)** If $a_1, a_2, a_3, \ldots a_n$ are the vertices in set *A* and $b_1, b_2, b_3, \ldots b_n$ are the vertices in set *B*, then one possible Hamilton circuit is $a_1, b_1, a_2, b_2, \ldots a_n, b_n, a_1$.

(b) Suppose $a_1, a_2, a_3, \ldots a_n$ are the vertices in set *A* and $b_1, b_2, b_3, \ldots b_{n+1}$ are the vertices in set *B*, then one possible Hamilton path is $b_1, a_1, b_2, a_2, \ldots b_n, a_n, b_{n+1}$.

(c) If $a_1, a_2, a_3, \ldots a_m$ are the vertices in set *A* and $b_1, b_2, b_3, \ldots b_n$ are the vertices in set *B*, then in any Hamilton path the *a*'s and *b*'s must alternate. This implies that either (i) there is the same number of *a*'s and *b*'s (*m* = *n*); (ii) there is one more of the *a*'s than there is of the *b*'s (*m* = *n* + 1); or (iii) there is one more of the *b*'s than there is of the *a*'s (*n* = *m* + 1). There are no other possibilities.

69. If $\deg(X) \geq N/2$ for every vertex *X*, then $\deg(X) + \deg(Y) \geq N/2 + N/2 = N$ for every pair of vertices *X* and *Y*.

Chapter 7

WALKING

7.1 Networks and Trees

1. No, the graph is disconnected. Vertices A, D, and G are connected to each other but not to the other four vertices. Looking at the graph one way, this may be difficult to see.

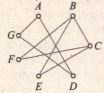

However, looking at a representation of the graph in another fashion, this is much easier to see.

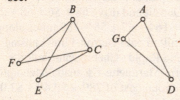

3. **(a)** 1; the shortest path connecting C and E has length 1 (they are connected to each other).

   (b) 3; the shortest path connecting A and E has length 3 (AB, BC, CE).

   (c) 4; the shortest path connecting A and F has length 4 (AB, BC, CE, EF).

5. **(a)** Several answers are possible. For example, vertices A and G have five degrees of separation (AB, BC, CE, EF, FG). So do vertices A and I (AB, BC, CE, EF, FI). Vertices B and H also have five degrees of separation (BC, CE, EF, FG, GH). So do K and H.

   (b) Only vertices A and H have six degrees of separation (AB, BC, CE, EF, FG, GH).

   (c) There are not any. A and H are two vertices farthest apart in the network (having six degrees of separation).

   (d) 6; the largest degree of separation between pairs of vertices is that between vertices A and H.

7. **(a)** N and J; O and J; O and K.
   Note that six is a large separation in this network so that choosing vertices that appear to be far apart is how one should approach this problem. Vertices N and O appear to be on one "end" of the network and vertices J and K appear to be on the other "end."

   (b) There are not any. The largest possible separation is between N (or O) and the cluster of vertices H, I, J, K, L, and M at the bottom right. To get to any of those vertices, one must go through H. That takes four links. From H it takes only one link to get to I, L, and M and two links to get to J and K.

   (c) 6; the largest degree of separation between pairs of vertices is that between vertices N and K (which is the same as that between vertices N and J, between O and J, and between O and K).

9. **(a)** 3; the shortest path connecting A and J is AB, BE, EJ.

   (b) 5; the shortest path connecting E and L is EB, BA, AD, DG, GL.

   (c) 8; the shortest path connecting M and P is MJ, JE, EB, BA, AD, DG, GL, LP.

   (d) 8; the largest degree of separation between A and any vertex in the top of this tree is 4 (from A to M, N, or O) and the largest degree of separation between A and any vertex in the bottom of this tree is also 4 (AD, DG, GL, LP).

11. (B) In a tree, the number of edges must be one less than the number of vertices. However, here the number of edges is one more than the number of vertices. So the network violates the N-1 edges property of trees.

13. (A) Since the network has one fewer edge than vertices, the network satisfies the $N - 1$ edges property.

15. (B) If $R = 1$, then $M = N$ (the number of edges is the same as the number of vertices). So, the network is not a tree since it violates the $N - 1$ edges property.

17. **(C)** We do not know if other pairs of vertices are connect by single or multiple paths. For example, the following network is a tree.

However, the following network, also having only one path connecting *A* and *J*, is not a tree.

19. **(B)** The network has 5 vertices and 10 edges (every edge produces two vertex degrees and there are 20 total degrees at the vertices). Since there are not 5 - 1 = 4 edges, the network violates the $N - 1$ edges property of trees.

21. **(A)** The network has 5 vertices and 4 edges (every edge produces two vertex degrees and there are 4 + 4 = 8 total degrees at the vertices). The network satisfies the $N - 1$ edges property of trees.

23. **(B)** The network has an Euler Circuit (see Section 5.3). Since it has a circuit, it cannot be a tree.

7.2 Spanning Trees, MSTs, and MaxSTs

25. **(a)** There are many spanning trees for this network. Below is one.

(b) Since this network has $N = 5$ vertices and $M = 10$ edges, the redundancy is $R = M - (N - 1) = 10 - (5 - 1) = 6$.

(c) 1; each vertex is connect to every other vertex by an edge.

27. **(a)** There is only one spanning tree for this network since it is a tree. The spanning tree is the tree itself.

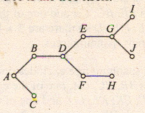

(b) Since this network has $N = 10$ vertices and $M = 9$ edges, the redundancy is $R = M - (N - 1) = 9 - (10 - 1) = 0$.

(c) 6; the shortest path connecting vertices *C* and *J*, for example, is *CA*, *AB*, *BD*, *DE*, *EG*, *GJ*. [Note: *C* and *I* are also separated by this degree.]

29. **(a)** Any one of the three edges *EI*, *DI*, or *DE* could be deleted to form a spanning tree.

(b) Any one of the four edges *DJ*, *EJ*, *EI*, or *DI* could be deleted to form a spanning tree.

(c) 6; Any one of the edges *DI*, *IJ*, *JE*, *EK*, *KL*, or *LD* could be deleted to form a spanning tree.

31. (a) Each spanning tree excludes one of the edges *AB, BC, CA* and one of the edges *DI, IE, EJ, JD* so there are $3 \times 4 = 12$ different spanning trees.

(b) Each spanning tree excludes one of the edges *AB, BC, CA*, one of the edges *DI, IJ, JE, EK, KL, LD*, and one of the edges *FG, GH, HF*, so there are $3 \times 6 \times 3 = 54$ different spanning trees.

33. (a) $6 + 2 \times 4 = 14$;

The key is to consider two cases: either *DF* is part of a spanning tree or it isn't. There are 6 spanning trees that do not contain edge *DF*. They can be found by removing *DF* along with one of the 6 edges of the circuit *A, B, C, D, E, F, A*.

There are 8 spanning trees that do contain edge *DF*. They can be found by removing one of the 4 edges other than *DF* of the circuit *A, B, C, D, F, A* along with one of the 2 edges (again, not *DF*) of the circuit *D, E, F, D*. Since there are 4 ways to do the former and 2 ways to do the latter, there are a total of $4 \times 2 = 8$ spanning trees containing the edge *DF*.

So, the number of different spanning trees is $6 + 8 = 14$.

(b) Certainly the edge *FE* should not be included. Not including edge *FD* also lengthens the path from *H* to *G*.

(c) Several answers are possible. The key idea is to be sure that edge *FE* is included so that the degree of separation between *H* and *G* is only 3.

7.3 Kruskal's Algorithm

35. A spanning tree for the 20 vertices will have exactly 19 edges and so the cost will be

$$19 \text{ edges} \times \frac{1 \text{ mile}}{2 \text{ edges}} \times \frac{\$40,000}{1 \text{ mile}} = \$380,000 .$$

37. Add edges to the tree in the following order:
EC, AD, AC, BC.

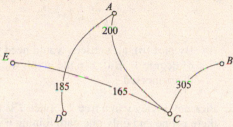

Weight = 165 + 185 + 200 + 305 = 855

39. Add edges to the tree in the following order:
DC, EF, EC, AB, AC.

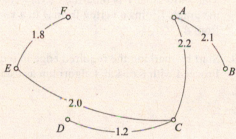

Weight = 1.2 + 1.8 + 2.0 + 2.1 + 2.2 = 9.3

41. Add edges to the tree in the following order:
AB, BD, BE, CD.

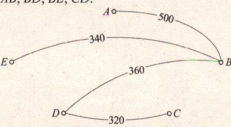

Weight = 500 + 360 + 340 + 320 = 1520

43. Add edges to the tree in the following order:
DF, BD, AE, BE, BC. Note that edges such as
BF, DE, and *AD* are not added to the MaxST
since they would form a circuit. Instead, edge
BC is the best you can do to involve vertex *C*.

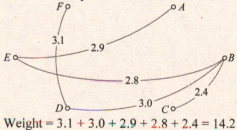

Weight = 3.1 + 3.0 + 2.9 + 2.8 + 2.4 = 14.2

JOGGING

45. Add edges to the tree in the following order:
Kansas City – Tulsa (248), Pierre – Minneapolis
(394), Minneapolis – Kansas City (447), Atlanta –
Columbus (533), Columbus – Kansas City (656).

47. (a) $k = 5, 2, 1, 0$ are all possible.
The network could have no circuits ($k=5$),

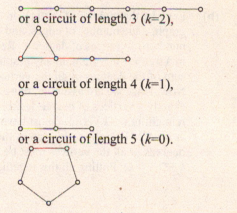

or a circuit of length 3 ($k=2$),

or a circuit of length 4 ($k=1$),

or a circuit of length 5 ($k=0$).

(b) Using the same pattern as in (a), values of
$k = 123$ and $0 \le k \le 120$ are all possible.

49. At each step of Kruskal's algorithm we choose
(from the available edges that do not close
circuits) the edge of least weight. When there
is only one choice at each step, there is only
one MST possible. The same argument
applies to a MaxST.

51. Step 1 of Kruskal's algorithm (which always
produces an MST) will always choose edge
XY. There is no other possibility when the
value of edge *XY* is a unique minimum.

53. Since the graph has no circuits, each
component has to be a tree. Thus, in each
component the number of edges is one less
than the number of vertices. It follows that $M = N - K$.

RUNNING

55. (a) $2N - 2$

A tree with N vertices has $N - 1$ edges, and, in any graph, the sum of the degrees of all the vertices is twice the number of edges.

(b) Let v be the number of vertices in the graph, e the number of edges, and k the number of vertices of degree 1. Recall that in a tree $v = e + 1$ and in any graph the sum of the degrees of all the vertices is $2e$. Now, since we are assuming there are exactly k vertices of degree 1, the remaining $v - k$ vertices must have degree at least 2. Therefore, the sum of the degrees of all the vertices must be at least $k + 2(v - k)$. Putting all this together we

have
$$2e \geq k + 2(v - k) = k + 2(e + 1 - k),$$
$$2e \geq k + 2e + 2 - 2k,$$
$$k \geq 2.$$

(c) By part (b), all vertices would need to be of degree 1. But then the graph would be disconnected.

57. Pick any vertex v in a tree T. Since T is a tree, there is one and only one path joining v with any other given vertex. So, let A be the set of vertices in T that are joined to v by a path of even length and let B be the set of vertices in T that are joined to v by a path of odd length. This shows that T is bipartite. Every edge of the graph T joins a vertex from A to a vertex from B.

59. Start by marking the required edge in red. Proceed with Kruskal's algorithm as usual.

Chapter 8

WALKING

8.2   Directed Graphs

1. **(a)** indeg(A) = 3; outdeg(A) = 2; there are three arcs, denoted by arrows, that come into vertex A (these are from vertices B, C, and E) and two arcs, once again denoted by arrows, that leave vertex A (for vertices B and C).

    (b) indeg(B) = 2; outdeg(B) = 2

    (c) indeg(D) = 3; outdeg(D) = 0

    (d) $3 + 2 + 1 + 3 + 1 = 10$; indeg(A) = 3, indeg(B) = 2, indeg(C) = 1, indeg(D) = 3, indeg(E) = 1.
    Note that the sum of the indegrees matches the number of arcs in the digraph.

    (e) $2 + 2 + 4 + 0 + 2 = 10$; outdeg(A) = 2, outdeg(B) = 2, outdeg(C) = 4, outdeg(D) = 0, outdeg(E) = 2.
    Note that the sum of the outdegrees is the same as the sum of the indegrees and also matches the number of arcs in the digraph.

3. **(a)** We look for paths from A to D that travel through exactly one other vertex (so that they are of length 2). To that end, A, B, D is one path. A, C, D is a second path. Traveling through vertex E will not work as there is no arc connecting vertex A to vertex E. Thus exactly two paths of length 2 from A to D exist: A, B, D and A, C, D.

    (b) In this situation, we look for all paths that start at A and end at D that go through exactly two of the vertices B, C, and E.
    - If we start at A and travel to B, we cannot get to either C or E with an arc, so that will not work.
    - If we start at A and travel to C, we have two choices to create a path of length 3. We can go to B and then D (creating path A, C, B, D) or we can go to E and then D (creating path A, C, E, D).
    - If we start at A and travel to E, we cannot get to either B or C with an arc, so that will not work.
    - Since there is no arc from A to D, we cannot take advantage of the fact that A connects to B which also connects back to A.
    Thus exactly two paths of length 3 from A to D exist: A, C, B, D and A, C, E, D.

    (c) There are a two ways to look for a path of length 4 from A to D: either find a path that goes through each vertex exactly once (which doesn't exist) or find a path that goes through one vertex (such as A) more than once. This later approach will work to produce such a path: A, B, A, C, D. Note: Another possible path constructed in this same way would be A, C, A, B, D.

    (d) As in (c), we seek a path that goes through one vertex (such as A) more than once. This time, however, we also seek a path that not only goes through A twice, but also goes through two of the vertices in the "middle" (namely B, C, and E). Two paths that do this travel from A to B and then back to A before moving on to D. They are A, B, A, C, B, D and A, B, A, C, E, D. A third path that does this travels from A to C and then to B before going back to A and then moving along to vertex D. It is A, C, B, A, B, D.

    (e) Since the outdegree of D is 0, there are no paths from D to A (no matter the length).

5. **(a)** A, B, A and A, C, A; Remember that cycle is a sequence of distinct adjacent arcs that starts and ends at the same vertex.

    (b) In this situation, such a cycle must begin with the arc AC (starting with arc AB fails – it leads to vertex D which has outdegree of 0). By starting with arc AC, there are two cycles having length 3 – A, C, B, A and A, C, E, A. [Again, going to vertex D from C leads nowhere since D has outdegree of 0.]

(c) Much like part (b), any such cycle must begin with either arc *AC* or arc *AB*. Since vertex *D* cannot be part of any cycle, we only need to investigate if vertex *E* could be included or not. Since there is no arc from *B* to *C*, the only cycles of length 4 cannot include vertex *E*. These cycles are either *A, B, A, C, A* or *A, C, A, B, A*.

(d) To form a cycle of length 5, we can (and must) take the two 2-cycles found in part (c) and grow one of the 2-cycles to a 3-cycle. In particular, we grow the 2-cycle *A, C, A* to *A, C, E, A*. This gives two possible cycles of length 5 – *A, C, E, A, B, A* and *A, B, A, C, E, A*.

7. (a) *C* and *E;* remember that given an arc *XY*, *X* is incident to *Y*. So *CA* and *EA* being arcs in this digraph suggest *C* and *E* are incident to *A*.

(b) *B* and *C;* given an arc *XY*, *Y* is incident from *X*. So *AB* and *AC* being arcs in this digraph suggest *B* and *C* are incident from *A*.

(c) *B, C,* and *E*

(d) No vertices are incident from *D*.

(e) *CD, CE* and *CA*

(f) No arcs are adjacent to *CD*.

9. (a) (b) 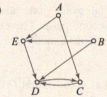 (c)

11. (a) 2 (these correspond to arcs *AB* and *AE*)

(b) 1 (this corresponds to arc *EA*)

(c) 1 (this corresponds to arc *DB*)

(d) 0 (there are no arcs incident to *D*)

13. (a) *A, B, D, E, F* is one possible path.

(b) *A, B, D, E, C, F*

(c) *B, D, E, B*

(d) The outdegree of vertex *F* is 0, so it cannot be part of a cycle.

(e) The indegree of vertex *A* is 0, so it cannot be part of a cycle.

(f) *B, D, E, B* is the only cycle

15.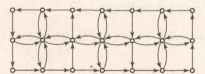

17. (a) *B*, since *B* is the only person that everyone respects.

(b) *A*, since *A* is the only person that no one respects.

(c) The individual corresponding to the vertex having the largest indegree (*B* in this example) would be the most reasonable choice for leader of the group since they would be the most respected person.

19. (a) The band's website and the TicketMonster Web site are likely to have large indegree and zero or very small outdegree (the local Smallville Web sites are likely to have hyperlinks coming from either). This makes *X* and *Y* the two likely choices. It is much more likely that the rock band's Web site will have a hyperlink to the ticket-selling Web site than the other way around. Give that there is a hyperlink from *Y* to *X* but not the other way, the most likely choice is *Y* for the rock band and *X* for the TicketMonster Website.

(b) *X*; See above.

(c) The radio station is likely to be a major source of information on anything related to the concert (the band, where to buy tickets, where to stay if you are from out of town, etc.), so it should be a vertex with high outdegree. *V* is the most likely choice.

(d) *Z* is a vertex to only the band's Web site and to the ticket selling Web site. None of the other Web sites are paying any attention to *Z*. This is most likely Joe Fan's blog.

(e) The two sister hotels would clearly have links to each other's Web sites, and since they are offering a special package for the concert, they are likely to have a link to the ticket-selling Web site. The logical choices are *U* and *W*.

21. The vertex set is {*START, A, B, C, D, E, F, G, H, END*} and the arc-set is {*START-A, START-H, AF, AC, B-END, CB, CE, D-END, E-END, FB, GB, GD, HG*}.

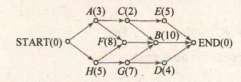

23. Vertex *A* represents the experiment taking 10 hours to complete. Vertices *B* and *C* represent the experiments requiring 7 hours each to complete. Vertices *D* and *E* represent the experiments requiring 12 hours each to complete. Vertices *F*, *G*, and *H* represent the experiments requiring 20 hours to complete.

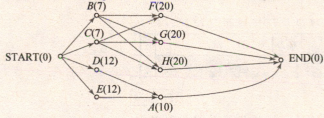

25.

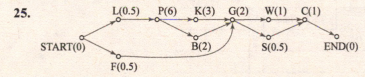

27.

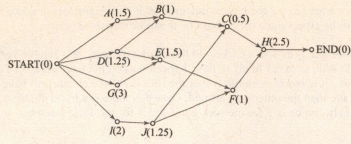

8.3 Priority-List Scheduling

29. (a) There are $31 \times 3 = 93$ processor hours available. Of these, $10 + 7 + 11 + 8 + 9 + 5 + 3 + 6 + 4 + 7 + 5 = 75$ hours are used. So, there must be $93 - 75 = 18$ hours of idle time.

 (b) There is a total of 75 hours of work to be done. Three processors working without any idle time would take $\dfrac{75}{3} = 25$ hours to complete the project.

31. There is a total of 75 hours of work to be done. Dividing the work equally between the six processors would require each processor to do $\dfrac{75}{6} = 12.5$ hours of work. Since there are no half hour jobs, the completion time could not be less than 13 hours.

33. ~~AB(8)~~ ~~AV(6)~~ ~~AF(5)~~ ~~IP(8)~~ ~~AP(7)~~ ~~IW(7)~~ ID(5), IP(4), ~~PS(4)~~ PU(3), HU(4), IC(1), PD(3), EU(2), FW(6)

35. (a) No. *B* (with processing time of 7) must be completed before *C* can be started.

 (b) No. *B* (with processing time of 7) must be completed before *D* can be started.

 (c) No. *G* is a ready task and it is ahead of *H* in the priority list. [Note: *C*, *D*, *E*, and *F* all must wait for *B* to be completed before they may be assigned.]

37.

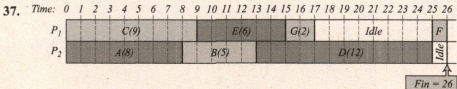

39.

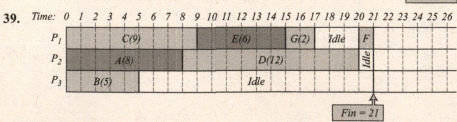

41. Both priority lists have all tasks in the top two paths of the digraph listed before any task in the bottom path.

8.4   The Decreasing-Time Algorithm

43. Decreasing-Time List: *D(12), C(9), A(8), E(6), B(5), G(2), F(1)*

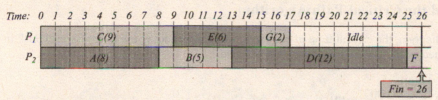

45. Decreasing-Time List: *K(20), D(15), I(15), C(10), E(7), H(5), J(5), B(4), G(4), F(3), A(2)*

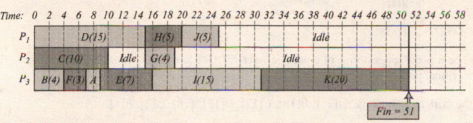

47. (a) Decreasing-Time List: *E(2), F(2), G(2), H(2), A(1), B(1), C(1), D(1)*

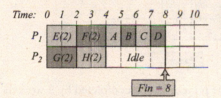

(b)

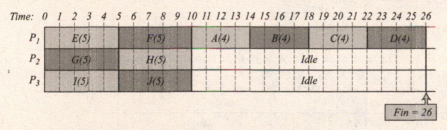

(c) The relative error expressed as a percent is $\varepsilon = \dfrac{Fin - Opt}{Opt} = \dfrac{8-6}{6} = 33\dfrac{1}{3}\%$.

49. (a) Decreasing-Time List: *E(5), F(5), G(5), H(5), I(5), J(5), A(4), B(4), C(4), D(4)*

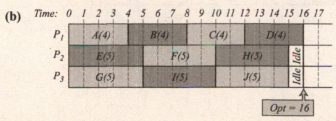

(b)

(c) The relative error expressed as a percent is $\varepsilon = \dfrac{Fin - Opt}{Opt} = \dfrac{26 - 16}{16} = 62.5\%$.

8.5 Critical Paths and the Critical-Path Algorithm

51. (a) As shown in the diagram, the critical time of each task is found by adding the times for each task in the shortest path formed between (and including) that task and *END*.

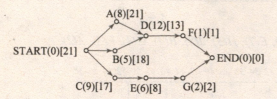

(b) The critical path is the longest time of a path between START and END. In this example, Start, A, D, F, END is the critical path, with a critical time of 21.

(c) From (a), the critical path list is: $A[21]$, $B[18]$, $C[17]$, $D[13]$, $E[8]$, $G[2]$, $F[1]$.

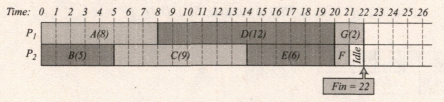

(d) There are a total of 43 work units, so the shortest time the project can be completed by 2 workers is $\dfrac{43}{2} = 21.5$ time units. Since there are no tasks with less than 1 time unit, the shortest time the project can actually be completed is 22 hours.

53. (a) The critical path list is: $B[46]$, $A[44]$, $E[42]$, $D[39]$, $F[38]$, $I[35]$, $C[34]$, $G[24]$, $K[20]$, $H[10]$, $J[5]$. So, the critical path is *START, B, E, I, K, END*.

(b)

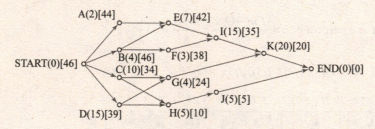

55. The project digraph, with critical times listed, is shown below.

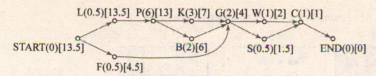

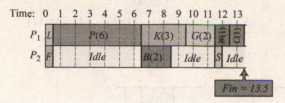

The critical-time priority list is $L[13.5]$, $P[13]$, $K[7]$, $B[6]$, $F[4.5]$, $G[4]$, $W[2]$, $S[1.5]$, $C[1]$ and the critical path is START[13.5], $L[13.5]$, $P[13]$, $K[7]$, $G[4]$, $W[2]$, $C[1]$, END[0].

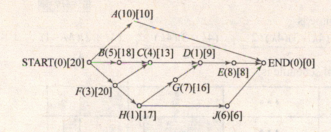

Since processing times are given in hours, the project finishing time is 13.5 hours.

57. (a) The critical path is START[20], $F[20]$, $H[17]$, $G[16]$, $D[9]$, $E[8]$, END[0]. The critical time is 20.

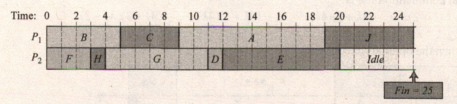

(b) The critical-time priority list is $F[20]$, $B[18]$, $H[17]$, $G[16]$, $C[13]$, $A[10]$, $D[9]$, $E[8]$, $J[6]$.

(c) The project finishing time is $5 + 4 + 10 + 6 = 25$.

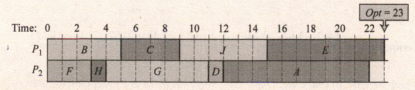

(d) An optimal schedule for $N = 2$ processors with finishing time $Opt = 23$ is shown below.

(e) The relative error expressed as a percent is

$$\varepsilon = \frac{Fin - Opt}{Opt} = \frac{25 - 23}{23} \approx 8.7\% .$$

JOGGING

59. Every arc of the graph contributes 1 to the indegree sum and 1 to the outdegree sum.

61. Assuming that all of the tasks are independent, then A, B, C, E, G, H, D, F, I is one possible priority list.

RUNNING

63. Consider four cases: $M = 4k$, $4k+1$, $4k+2$, and $4k+3$ for some positive integer k.
If $M = 4k$, then the optimal schedule is

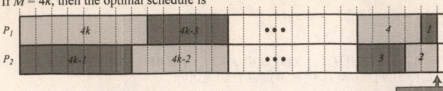

and the optimal completion time is $\dfrac{1+2+3+\ldots+4k}{2} = \dfrac{(4k+1)(4k)/2}{2} = \dfrac{(4k+1)(4k)}{4} = \dfrac{(M+1)(M)}{4}$.

If $M = 4k+1$, then the optimal schedule is

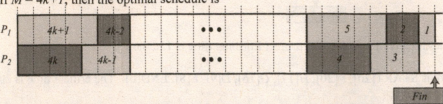

and the optimal completion time is
$1 + \dfrac{2+3+\ldots+(4k+1)}{2} = 1 + \dfrac{(4k+3)(4k)/2}{2} = 1 + \dfrac{(4k+3)(4k)}{4} = 1 + \dfrac{(M+2)(M-1)}{4}$.

If $M = 4k+2$, then the optimal schedule is

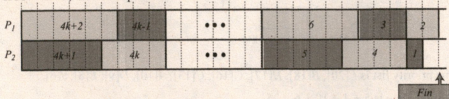

and the optimal completion time is
$1 + 2 + \dfrac{3+\ldots+(4k+2)}{2} = 3 + \dfrac{(4k+5)(4k)/2}{2} = 3 + \dfrac{(4k+5)(4k)}{4} = 3 + \dfrac{(M+3)(M-2)}{4}$.

If $M = 4k+3$, then the optimal schedule is

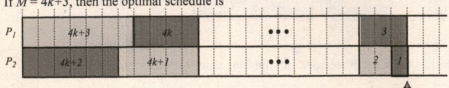

and the optimal completion time is $\dfrac{1+2+3+\ldots+(4k+3)}{2} = \dfrac{(4k+4)(4k+3)/2}{2} = \dfrac{(M+1)(M)}{4}$.

65. Consider the case in which $M = 6$. In that case, $Opt = 64$. Since all tasks are independent, the critical-time priority list is identical to a decreasing-time list.

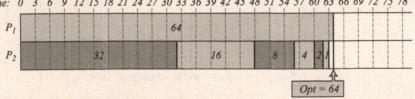

In general, $Opt = 2^M$ and the schedule is similar.

Chapter 9

WALKING

9.1. Sequences and Population Sequences

1. (a) $A_1 = 1^2 + 1 = 2$

(b) $A_{100} = 100^2 + 1 = 10,001$

(c) If $A_N = 10$, then $N^2 + 1 = 10$. So $N^2 = 9$ and $N = \pm 3$. Since the subscripts N are all integers satisfying $N \geq 1$ in this sequence, we have $N = 3$.

3. (a) $A_1 = \dfrac{4 \times 1}{1+3} = \dfrac{4}{4} = 1$

(b) $A_9 = \dfrac{4 \times 9}{9+3} = \dfrac{36}{12} = 3$

(c) If $A_N = \dfrac{5}{2}$, then $\dfrac{4N}{N+3} = \dfrac{5}{2}$. So $8N = 5(N+3)$ and $8N = 5N + 15$. This leads to $3N = 15$ and $N = 5$.

5. (a) $A_1 = (-1)^{1+1} = (-1)^2 = 1$

(b) $A_{100} = (-1)^{100+1} = (-1)^{101} = -1$

(c) If $A_N = 1$, then $(-1)^{N+1} = 1$. This is true exactly when $N+1$ is an even integer (Since $N \geq 1$, this means 2, 4, 6, 8,…). But $N+1$ is a positive even integer exactly when N is a positive odd integer. That is, when $N = 1, 3, 5, 7, …$

7. (a) $A_3 = 2A_{3-1} + A_{3-2} = 2A_2 + A_1 = 2(1) + 1 = 3$
$A_4 = 2A_{4-1} + A_{4-2} = 2A_3 + A_2 = 2(3) + 1 = 7$
$A_5 = 2A_{5-1} + A_{5-2} = 2A_4 + A_3 = 2(7) + 3 = 17$
$A_6 = 2A_{6-1} + A_{6-2} = 2A_5 + A_4 = 2(17) + 7 = 41$

(b) We could continue as in (a).
$A_7 = 2A_6 + A_5 = 2(41) + 17 = 99$
$A_8 = 2A_7 + A_6 = 2(99) + 41 = 239$
Or, we could think using the rule "double the last term in the sequence and add the term prior to that." That is, double the $(N-1)^{st}$ term and add the $(N-2)^{nd}$ term. Using this rule, A_7 would be twice 41 added to 17 (or 99). Then A_8 would be twice 99 added to 41 (or 239).

9. (a) $A_3 = A_{3-1} - 2A_{3-2}$
$= A_2 - 2A_1$
$= -1 - 2(1)$
$= -3$
$A_4 = A_{4-1} - 2A_{4-2}$
$= A_3 - 2A_2$
$= -3 - 2(-1)$
$= -1$
$A_5 = A_{5-1} - 2A_{5-2}$
$= A_4 - 2A_3$
$= -1 - 2(-3)$
$= 5$
$A_6 = A_{6-1} - 2A_{6-2}$
$= A_5 - 2A_4$
$= 5 - 2(-1)$
$= 7$

(b) We could continue as in (a).
$A_7 = A_6 - 2A_5 = 7 - 2(5) = -3$
$A_8 = A_7 - 2A_6 = -3 - 2(7) = -17$
Or, we could think using the rule "from the last term in the sequence, subtract twice the term before that." That is, from the $(N-1)^{st}$ term, subtract twice the $(N-2)^{nd}$ term. Using this rule, A_7 would be 7 minus twice 5 (or -3). Then A_8 would be -3 minus twice 7 (or -17).

11. (a) There are (at least) two different ways to observe the pattern here. The first is to notice that these are perfect squares (of integers). So the pattern can be recognized as $1^2 = 1$, $2^2 = 4$, $3^2 = 9$, $4^2 = 16$, $5^2 = 25$, $6^2 = 36$, $7^2 = 49, …$
Another way to observe the pattern is to notice that consecutive odd integers are being added. So the pattern could also be recognized as 1, $1+3 = 4$, $4+5 = 9$, $9+7 = 16$, $16+9 = 25$, $25+11 = 36$, $36+13 = 49, …$
Either way, the next two terms could reasonably be listed as 36 and 49.

(b) Recognizing the pattern (see (a)) by
$A_1 = 1^2 = 1$, $A_2 = 2^2 = 4$, $A_3 = 3^2 = 9$,
$A_4 = 4^2 = 16$, $A_5 = 5^2 = 25$, …we can explicitly describe the sequence using $A_N = N^2$. [Recognize the pattern as the subscript being squared.]

(c) Recognizing the pattern (see (a)) by
$P_0 = 1^2$, $P_1 = 2^2$, $P_2 = 3^2$, $P_3 = 4^2$,
$P_4 = 5^2$, …we can explicitly describe the sequence using $P_N = (N+1)^2$.
[Recognize the pattern as the square of one more than the subscript being squared.]

13. (a) There are (at least) two different ways to observe the pattern here. The first is to notice that consecutive integers are being added. So the pattern could be recognized as 1, $1+2=3$, $3+3=6$, $6+4=10$, $10+5=15$, $15+6=21$, $21+7=28$, …
Another way to observe the pattern is to look at double each value in the sequence: 0, 2, 6, 12, 20, 30, 42, … The pattern here can be recognized as $0 \times 1 = 0$, $1 \times 2 = 2$, $2 \times 3 = 6$, $3 \times 4 = 12$, $4 \times 5 = 20$, $5 \times 6 = 30$, $6 \times 7 = 42$, $7 \times 8 = 56$, … So cutting these each in half gives the pattern $(0 \times 1)/2 = 0$, $(1 \times 2)/2 = 1$, $(2 \times 3)/2 = 3$, $(3 \times 4)/2 = 6$, $(4 \times 5)/2 = 10$, $(5 \times 6)/2 = 15$, $(6 \times 7)/2 = 21$, $(7 \times 8)/2 = 28$, … Either way, the next two terms could reasonably be listed as 21 and 28.

(b) Recognizing the pattern (see (a)) by
$A_1 = (0 \times 1)/2 = 0$, $A_2 = (1 \times 2)/2 = 1$,
$A_3 = (2 \times 3)/2 = 3$, $A_4 = (3 \times 4)/2 = 6$,
$A_5 = (4 \times 5)/2 = 10$, $A_6 = (5 \times 6)/2 = 15$,
…we can explicitly describe the sequence using $A_N = ((N-1) \times N)/2$. [The pattern is half the product of one less than the subscript multiplied by the subscript.]

(c) Recognizing the pattern (see (a)) by
$P_0 = (0 \times 1)/2 = 0$, $P_1 = (1 \times 2)/2 = 1$,
$P_2 = (2 \times 3)/2 = 3$, $P_3 = (3 \times 4)/2 = 6$,
$P_4 = (4 \times 5)/2 = 10$, $P_5 = (5 \times 6)/2 = 15$, …
we can explicitly describe the sequence using $P_N = (N(N+1))/2$.
[Recognize the pattern as half the product of the subscript multiplied by one more than the subscript.]

15. (a) First, it helps to list the sequence entirely in terms of improper fractions to spot a pattern: $\frac{4}{4}, \frac{8}{5}, \frac{12}{6}, \frac{16}{7}, \frac{20}{8}, \ldots$ At this point, we then notice that the numerators are integer multiples of 4 and the denominators are consecutive integers. So the pattern could be seen to continue as $\frac{24}{9}, \frac{28}{10}, \ldots$

(b) Recognizing the pattern (see (a)) by
$A_1 = \frac{1 \times 4}{4}$, $A_2 = \frac{2 \times 4}{5}$, $A_3 = \frac{3 \times 4}{6}$,
$A_4 = \frac{4 \times 4}{7}$, $A_5 = \frac{5 \times 4}{8}$, $A_6 = \frac{6 \times 4}{9}$,
$A_7 = \frac{7 \times 4}{10}$, … we can explicitly describe the sequence using $A_N = \frac{N(4)}{N+3} = \frac{4N}{N+3}$.
[The pattern is the subscript times four divided by three more than the subscript.]

17. (a) We may recognize the pattern as similar to that found in Exercise 12: $\frac{1}{2!}, \frac{1}{3!}, \frac{1}{4!}, \frac{1}{5!}, \ldots$ The next term, which represents the probability that six passengers will board in order of decreasing seat numbers is $\frac{1}{6!}$ or $\frac{1}{720}$.

(b) Recognizing the pattern by $A_1 = \dfrac{1}{2!}$,

$$A_2 = \frac{1}{3!}, \quad A_3 = \frac{1}{4!}, \quad A_4 = \frac{1}{5!}, \quad A_5 = \frac{1}{6!}, \dots$$

we can explicitly describe the sequence

using $A_N = \dfrac{1}{(N+1)!}$. Since A_N represents the probability that $N+1$ passengers board in order of decreasing seat numbers, we seek the value of A_{11}. The desired

probability is $A_{11} = \dfrac{1}{(12)!} = \dfrac{1}{479{,}001{,}600}$.

9.2. The Linear Growth Model

19. (a) $P_1 = P_0 + 125 = 80 + 125 = 205$;
$P_2 = P_1 + 125 = 205 + 125 = 330$;
$P_3 = P_2 + 125 = 330 + 125 = 455$

(b) Notice the pattern to construct an explicit formula.
$P_1 = 80 + 125$;
$P_2 = 80 + 2(125)$;
$P_3 = 80 + 3(125)$;
$P_4 = 80 + 4(125) \dots$
It follows that $P_N = 80 + N(125)$ which can also be written as $P_N = 80 + 125N$.

(c) $P_{100} = 80 + 100(125) = 12{,}580$

21. (a) $P_1 = P_0 - 25 = 578 - 25 = 553$;
$P_2 = P_1 - 25 = 553 - 25 = 528$;
$P_3 = P_2 - 25 = 528 - 25 = 503$

(b) Notice the pattern to construct an explicit formula.
$P_1 = 578 - 25$;
$P_2 = 578 - 2(25)$;
$P_3 = 578 - 3(25)$;
$P_4 = 578 - 4(25) \dots$
It follows that $P_N = 578 - N(25)$ which can also be written as $P_N = 578 - 25N$.

(c) $P_{23} = 578 - 23(25) = 3$

23. (a) Here we can use the fact that
$P_{10} = P_0 + 10d$ where d is the common difference between terms of the sequence.

So $38 = 8 + 10d$ and $30 = 10d$. It follows that the common difference is $d = 3$. So
$P_1 = 8 + 3$;
$P_2 = 8 + 2(3)$;
$P_3 = 8 + 3(3)$;
$P_4 = 8 + 4(3) \dots$ It follows that
$P_N = 8 + N(3)$ which can also be written as $P_N = 8 + 3N$.

(b) $P_{50} = 8 + 50(3) = 158$ (using the explicit formula found in (a)).

25. (a) Model the unemployment rates using an arithmetic sequence with $d = -0.2\%$ where $P_0 = 8.9\%$, $P_1 = 8.7\%$, $P_2 = 8.5\%$, $P_3 = 8.3\%$ and, in general, P_N represents the official unemployment rate after October of 2011. The unemployment rate in January would then be the value of P_{15} (since it's the 15th month after October of 2011).
$P_{15} = 8.9\% + 15(-0.2\%) = 5.9\%$

(b) An explicit linear model for the unemployment rate N months after October 2011 can be given
by $P_N = 8.9\% + N(-0.2\%)$. To determine when the U.S. would reach a zero unemployment rate, we determine the first N value for which $P_N \leq 0$. But
$8.9\% + N(-0.2\%) \leq 0$ when
$-0.2\%N \leq -8.9\%$. That is, when
$N \geq \dfrac{-8.9\%}{-0.2\%}$. The smallest integer value
of N for which this happens is $N = 45$. July of 2015 is the 45th month after October of 2011. The unemployment rate for June, 2015, would be 0.1% and would, in theory, hit 0% during July of that year.

27. (a) The year 2012 is 17 years after 1995. So a person born in 2012 corresponds to a value of $N = 17$ in this model. The life expectancy of such a person is
$L_{17} = 66.17 + 0.96(17) = 82.49$ years.

(b) We solve $L_N = 66.17 + 0.96N = 90$ for N. But $0.96N = 23.83$ gives $N \approx 24.82$ years. The year 2020 is the 25th year after 1995 and the life expectancy in that year would be $L_{25} = 66.17 + 0.96(25) \approx 90.17$ years.

29. The common difference in this arithmetic sequence is $d = 5$. We also have $P_0 = 2$ and $P_{99} = 497$. The sum of 100 terms in this arithmetic sequence is then

$$2 + 7 + 12 + \ldots + 497 = \frac{(2 + 497) \times 100}{2}$$
$$= 24,950.$$

31. (a) If 309 is the Nth term of the sequence,

$$309 = 12 + 3(N - 1)$$
$$309 = 12 + 3N - 3$$
$$300 = 3N$$
$$N = 100$$

So, 309 is the 100th term of the sequence.

(b) $12 + 15 + 18 + \ldots + 309 = \dfrac{(12 + 309) \times 100}{2}$
$$= 16,050.$$

33. (a) The odd numbers form an arithmetic sequence with common difference $d = 2$. We first find the number of terms in the sum $1 + 3 + 5 + 7 + \ldots + 149$. Writing $149 = 1 + 2(N - 1) = 1 + 2N - 2 = 2N - 1$ and solving for N gives $N = 75$. We then use the arithmetic sum formula to find

$$1 + 3 + 5 + 7 + \ldots + 149 = \frac{(1 + 149) \times 75}{2} = 5625.$$

(b) This is a sum of 100 terms in the same arithmetic sequence as part (a). The 100th term is given by $1 + 2(100 - 1) = 199$. We use the arithmetic sum formula to find

$$1 + 3 + 5 + \ldots + 199 = \frac{(1 + 199) \times 100}{2} = 10,000.$$

35. (a) $P_{38} = 137 + 38(2) = 213$

(b) $P_N = 137 + 2N$

(c) $137 \times \$1 \times 52 = \$7124$

(d) When just counting the newly installed lights, $P_0 = 0$ and $P_{51} = 2 \times 51 = 102$. (The lights installed in the 52nd week aren't in operation during the 52-week period.)

$$0 + 2 + 4 + \ldots + 102 = \frac{(0 + 102) \times 52}{2}$$
$$= 2652$$

The cost is $2,652.

9.3. The Exponential Growth Model

37. (a) $P_1 = 11 \times 1.25 = 13.75$

(b) $P_9 = 11 \times (1.25)^9 \approx 81.956$

(c) $P_N = 11 \times (1.25)^N$

39. (a) $P_1 = 4P_0 = 4 \times 5 = 20$;
$P_2 = 4P_1 = 4 \times 20 = 80$;
$P_3 = 4P_2 = 4 \times 80 = 320$

(b) Notice the pattern:
$P_1 = 5 \times 4$;
$P_2 = (5 \times 4) \times 4$;
$P_3 = (5 \times 4 \times 4) \times 4$;
$P_4 = (5 \times 4 \times 4 \times 4) \times 4$; ...
So, explicitly we have $P_N = 5 \times 4^N$.

(c) We determine the smallest value of N such that $P_N = 5 \times 4^N \geq 1,000,000$. That is, for which $4^N \geq 200,000$. Trial and error suggest the smallest integer value of N that does this is $N = 9$. After 9 generations, in fact, the population will be over 5.2 million.

41. (a) $P_N = 1.50 P_{N-1}$ where $P_0 = 200$
We multiply by 1.50 each year to account for an *increase* by 50%.

(b) $P_N = 200 \times 1.5^N$

(c) $P_{10} = 200 \times 1.5^{10} \approx 11,533$
A good estimate would be to say that about 11,500 crimes will be committed in 2020.

43. If r represents the growth rate, then the population sequence is described by P_0,
$P_1 = (1 + r)P_0$, $P_2 = (1 + r)^2 P_0$, ... where P_N represents the number of mathematics majors N years after 2010. We have $P_0 = 425$ and $P_1 = 463$. Solving $463 = (1 + r)(425)$ gives us a growth rate of $r = 38/425$. This is approximately 8.94%.

45. If r represents the growth rate, then the population sequence is described by P_0,

$P_1 = (1+r)P_0$, $P_2 = (1+r)^2 P_0$, ... where P_N represents the undergraduate enrollment at Bright State University N years after 2010. We have $P_0 = 19,753$ and $P_1 = 17,389$.

Solving $17,389 = (1+r)(19,753)$ gives us a growth rate of $r = -2364/19,753$. This is approximately -11.97%. [The growth rate being negative tells us that the population is decreasing in size.]

47. (a) $R = \dfrac{P_1}{P_0} = \dfrac{6}{2} = 3$

[Note that $R = \dfrac{P_2}{P_1} = \dfrac{18}{6}$ too.]

(b) The geometric sum formula:

$$P_0 + RP_0 + R^2 P_0 + \ldots + R^{N-1} P_0 = \frac{(R^N - 1)P_0}{R-1}$$

We have $P_0 = 2$, $R = 3$, and $N = 21$ (there are 21 terms being summed). The sum is $\dfrac{(3^{21} - 1)(2)}{3-1} = 10,460,353,202$.

49. (a) $R = \dfrac{P_1}{P_0} = \dfrac{2}{4} = 0.5$

[Note that $R = \dfrac{P_2}{P_1} = \dfrac{1}{2}$ too.]

(b) The geometric sum formula:

$$P_0 + RP_0 + R^2 P_0 + \ldots + R^{N-1} P_0 = \frac{(R^N - 1)P_0}{R-1}$$

We have $P_0 = 4$, $R = 0.5$, and $N = 12$ (there are 12 terms being summed). The sum is $\dfrac{((0.5)^{12} - 1)(4)}{0.5-1} \approx 7.998$.

51. (a) Observing the geometric sum formula,

$$P_0 + RP_0 + R^2 P_0 + \ldots + R^{N-1} P_0 = \frac{(R^N - 1)P_0}{R-1},$$

we have $P_0 = 1$, $R = 2$, and $N = 16$. That is, the first term is 1, the common ratio is 2, and there are $N = 16$ terms.

So the sum is

$$\frac{(R^N - 1)P_0}{R-1} = \frac{(2^{16} - 1) \cdot 1}{2-1} = 2^{16} - 1 = 65,535.$$

(b) This is as in (a) except we no longer have $N = 16$. So the sum is

$$\frac{(R^N - 1)P_0}{R-1} = \frac{(2^N - 1) \cdot 1}{2-1} = 2^N - 1.$$

9.4. The Logistic Growth Model

53. (a) $p_1 = 0.8 \times (1 - 0.3) \times 0.3 = 0.1680$

(b) $p_2 = 0.8 \times (1 - 0.168) \times 0.168 \approx 0.1118$

(c) $p_3 = 0.8 \times (1 - 0.11182) \times 0.11182$
≈ 0.07945
Thus, approximately 7.945% of the habitat's carrying capacity is taken up by the third generation.

55. (a) Using the formula $p_{N+1} = r(1 - p_N)p_N$ and a calculator with a memory register or a spreadsheet, we get
$p_1 = 0.1680$, $p_2 \approx 0.1118$, $p_3 \approx 0.0795$,
$p_4 \approx 0.0585$, $p_5 \approx 0.0441$, $p_6 \approx 0.0337$,
$p_7 \approx 0.0261$, $p_8 \approx 0.0203$, $p_9 \approx 0.0159$,
$p_{10} \approx 0.0125$.

(b) Since $p_N \to 0$ this logistic growth model predicts extinction for this population.

57. (a) Using the formula $p_{N+1} = r(1 - p_N)p_N$ and a calculator with a memory register or a spreadsheet, we get
$p_1 = 0.4320$, $p_2 \approx 0.4417$, $p_3 \approx 0.4439$,
$p_4 \approx 0.4443$, $p_5 \approx 0.4444$, $p_6 \approx 0.4444$,
$p_7 \approx 0.4444$, $p_8 \approx 0.4444$, $p_9 \approx 0.4444$,
$p_{10} \approx 0.4444$.

(b) The population becomes stable at $\dfrac{4}{9} \approx 44.44\%$ of the habitat's carrying capacity.

59. (a) $p_1 = 0.3570$, $p_2 \approx 0.6427$, $p_3 \approx 0.6429$,
$p_4 \approx 0.6428$, $p_5 \approx 0.6429$, $p_6 \approx 0.6428$,
$p_7 \approx 0.6429$, $p_8 \approx 0.6428$, $p_9 \approx 0.6429$,
$p_{10} \approx 0.6428$

(b) The population becomes stable at
$\frac{9}{14} \approx 64.29\%$ of the habitat's carrying capacity.

61. (a) $p_1 = 0.5200$, $p_2 = 0.8112$, $p_3 \approx 0.4978$,
$p_4 \approx 0.8125$, $p_5 \approx 0.4952$, $p_6 \approx 0.8124$,
$p_7 \approx 0.4953$, $p_8 \approx 0.8124$, $p_9 \approx 0.4953$,
$p_{10} \approx 0.8124$

(b) The population settles into a two-period cycle alternating between a high-population period at approximately 81.24% and a low-population period at approximately 49.53% of the habitat's carrying capacity.

JOGGING

63. (a) Exponential ($r = 2$)

(b) Linear ($d = 2$)

(c) Logistic

(d) Exponential ($r = \frac{1}{3}$)

(e) Logistic

(f) Linear ($d = -0.15$)

(g) Linear ($d = 0$), Exponential ($r = 1$), and/or Logistic (they all apply!)

65. The first N terms of the arithmetic sequence are
$P_0, P_0 + d, P_0 + 2d, \ldots, P_0 + (N-1)d$. Their sum is
$$\frac{(P_0 + [P_0 + (N-1)d]) \times N}{2} = \frac{N}{2}[2P_0 + (N-1)d].$$

67. Similar to Exercise 33(b), we compute $\underbrace{1 + 3 + 5 + \ldots}_{N \text{ terms}}$. The Nth term is given by
$1 + 2(N-1) = 1 + 2N - 2 = 2N - 1$. We use the arithmetic sum formula to find
$$1 + 3 + 5 + \ldots + 2N - 1 = \frac{(1 + 2N - 1) \times N}{2}$$
$$= \frac{2N \cdot N}{2}$$
$$= N^2.$$

69. One example is $P_0 = 8, R = \frac{1}{2}$.

Then $P_1 = 4, P_2 = 2, P_3 = 1, P_4 = \frac{1}{2}, P_5 = \frac{1}{4} \ldots$

RUNNING

71. No. This would require
$p_0 = p_1 = 0.8(1 - p_0)p_0$ and $p_0 = 0$ or
$1 = 0.8(1 - p_0)$. So, $p_0 = 0$ or $p_0 = -0.25$,
neither of which are possible.

73. If the population is constant, $p_1 = p_0$. Thus,
$rp_0(1 - p_0) = p_0$, i.e., $r(1 - p_0) = 1$. Solving
for p_0 gives $p_0 = \frac{r-1}{r}$.

75. (a) If $r > 4$, and $p_N = 0.5$,
$$p_{N+1} = r(1 - p_N)p_N$$
$$= 4(1 - 0.5)(0.5)$$
$$= 0.25r$$
$$> 1$$
This is impossible since the population cannot be more than 100% of the carrying capacity.

(b) If $r = 4$, and $p_N = 0.5$,
$p_{N+1} = r(1 - p_N)p_N = 4(1 - 0.5)(0.5) = 1$
and hence $p_K = 0$ for all $K > N+1$ (i.e. the population becomes extinct.)

(c) ¼
The graph of $(1 - p)p = -p^2 + p$ is an inverted parabola which crosses the x-axis at 0 and 1. Thus, its maximum value occurs when $p = \frac{1}{2}$.

(d) From part (c), if $0 < p < 1$,
$0 < (1 - p)p < \frac{1}{4}$ and so if we also have
$0 < r < 4$, then $0 < r(1 - p)p < 1$.
Consequently, if we start with $0 < p_0 < 1$
and $0 < r < 4$, then $0 < p_N < 1$ for every positive integer N.

Chapter 10

WALKING

10.1 Percentages

1. (a) 0.0625 (drop the % sign and move the decimal point two places to the left)

    (b) 0.0375 $\left[\text{Note that } 3\frac{3}{4}\% = 3.75\% \right]$

    (c) 0.007 $\left[\text{First write } \frac{7}{10}\% = 0.7\% \right]$

3. (a) 0.0082 (Remember, drop the % sign and move the decimal point two places to the left.)

    (b) 0.0005

5. The score for Lab 1 was $\frac{61}{75} = 81\frac{1}{3}\%$. The score for Lab 2 was $\frac{17}{20} = 85\%$. The score for Lab 3 was $\frac{118}{150} = 78\frac{2}{3}\%$. So a ranking would be Lab 2 (85%), Lab 1 ($81\frac{1}{3}\%$), Lab 3 ($78\frac{2}{3}\%$).

7. 215; If 14% of the pieces are missing, then 86% of the pieces are present. 86% of 250 is $0.86 \times 250 = 215$.

9. Suppose that x represents the tax rate on the earrings. Then, $(1+x)\$6.95 = \$7.61$. So, $x = \frac{\$7.61}{\$6.95} - 1 = 0.095$. It follows that the tax rate is 9.5%.

11. If T was the starting tuition, then the tuition at the end of one year is 110% of T. That is, $(1.10)T$. The tuition at the end of two years is 115% of what it was after one year. That is, after two years, the tuition was $(1.15)(1.10)T$. Likewise, the tuition at the end of the three years was $(1.10)(1.15)(1.10)T = 1.3915T$. The total percentage increase was 39.15%.

13. Suppose that the cost of the shoes is originally $\$C$. After the mark up, the price is $\$2.2C$. When the shoes go on sale, they are priced at 80% of that price or $\$(0.8)(2.2)C$. When the shoes find their way to the clearance rack, they are priced at 70% of the sale price which is $\$(0.7)(0.8)(2.2)C$. After the customer applies the coupon, the final price is 90% of the clearance price. That is, $\$(0.9)(0.7)(0.8)(2.2)C = \$1.1088C$. So, the profit on these shoes is 10.88% more than the original cost $\$C$. [If you still don't believe it, imagine the shoes are priced at an even \$100 originally and do the computations.]

15. Suppose that when the week began the DJIA had a value of A. The following chart describes the value at the end of each day of the week.

Day	DJIA value
Mon.	$(1.025)A$
Tues.	$(1.121)(1.025)A$
Wed.	$(0.953)(1.121)(1.025)A$
Thur.	$(1.008)(0.953)(1.121)(1.025)A$
Fri.	$(0.946)(1.008)(0.953)(1.121)(1.025)A$

At the end of the week the DJIA has a value of approximately $1.044A$, a net increase of 4.4%.

10.2 Simple Interest

17. $\$875(1 + 4 \times 0.0428) = \$1,024.80$

19. Suppose the purchase price is $\$P$. When you cash in the bond after four years, it is worth $\$P(1 + 4 \times 0.0575) = \$4920$. This is the face value. Solving for $\$P$, the purchase price for the bond is $\$P = \dfrac{\$4920}{(1 + 4 \times 0.0575)} = \$4000$.

21. Using the simple interest formula, we solve $\$5400(1 + 8 \cdot r) = \$8316$ for r to find $r = \dfrac{\$8316 - \$5400}{\$5400 \times 8} = 0.0675$. So, the APR is 6.75%.

23. If the principal doubles in 12 years, then a face value of $2P$ can be used in the simple interest formula. We solve $\$P(1+12\cdot r) = \$2P$ for r

to find $r = \dfrac{\$2P - \$P}{\$P \times 12} = \dfrac{1}{12} \approx 0.0833$. This

gives an APR of 8.33% (to the nearest hundredth percent).

25. We use the simple interest formula $I = Prt$ where $I = \$17$, $P = \$100$, and $t = 14/365$ years to solve $\$17 = \$100 \cdot r \cdot (14/365)$ for r. We find $r \approx 4.432$. So, the APR is 443.2%. [Yes, that is correct. Over 400%.]

10.3 Compound Interest

27. **(a)** $\$3250(1+0.09)^4 = \$4587.64$

(b) Since interest is credited to the account at the end of each year, no growth takes place during the last half year. The balance after this time is thus $\$3250(1+0.09)^5 = \$5000.53$.

29. $\$25,000(1+0.035)^{20} = \$49,744.72$

31. **(a)** The periodic interest rate (here $r = 0.03$ and $n = 12$) is

$$p = \frac{r}{n} = \frac{0.03}{12} = 0.0025 = 0.25\%.$$

(b) The future value of $\$1580$ in 3 years (36 months of growth at 0.25% interest each month) is $\$1580(1+0.0025)^{36} = \$1728.60$.

33. **(a)** The periodic interest rate (here $r = 0.0365$ and $n = 365$) is $p = \dfrac{r}{n} = \dfrac{0.0365}{365} = 0.0001$. That is, 0.01% per day!

(b) The future value of $\$1580$ in 3 years ($3 \times 365 = 1095$ days of growth) is

$$\$1580\left(1+\frac{0.0365}{365}\right)^{1095} \approx \$1762.83.$$

35. The future value F of P dollars compounded continuously for t years with APR r is $F = Pe^{rt}$. In this problem, we have $P = 1580$, $r = 0.03$, and $t = 3$. So, the future value is $F = \$1580e^{(0.03)\times 3} = \$1728.80$.

37. In this exercise we have $P = 1580$, $r = 0.0365$, and $t = \dfrac{500}{365}$ (years). So, the future value is $F = \$1580e^{(0.0365)\times\frac{500}{365}} \approx \$1661.008332$. Rounded to the nearest dollar, this is $\$1661$.

39. In this exercise, we know the future value $F = \$868.80$ and seek the original principal P. We also have $r = 0.0275$ and $t = 3$ (years). To determine the original principal P, we solve $\$868.80 = Pe^{(0.0275)\times 3}$ for P. Solving gives

$$P = \$\frac{868.80}{e^{(0.0275)\times 3}} \approx \$800.$$

41. **(a)** The APY is $\left(1+\dfrac{0.06}{1}\right)^1 - 1 = 0.06 = 6\%$.

(b) $\left(1+\dfrac{0.06}{2}\right)^2 - 1 = 0.0609 = 6.09\%$.

(c) $\left(1+\dfrac{0.06}{12}\right)^{12} - 1 \approx 0.061678 = 6.1678\%$

(d) $1e^{0.06\times 1} - 1 \approx 0.061837 = 6.1837\%$

43. We estimate the value of t that makes $\$2P = P(1+0.06)^t$ true. That is, we estimate t that solves $2 = (1.06)^t$. Trial and error gives $t = 12$ years.

10.4 Retirement Savings

45. **(a)** We use the Retirement Savings Formula (with annual contributions),

$$\$V = \$P\left[\frac{(1+r)^Y - 1}{r}\right], \text{ where the annual}$$

contribution (in dollars) made at the end of each year is $P = 1500$, the annual interest rate is $r = 0.066$, and the number of years of retirement savings is $Y = 45$. The value of the retirement savings account is then calculated to be

$$\$V = \$1500\left[\frac{(1+0.066)^{45} - 1}{0.066}\right]$$

$$\approx \$380,566.52.$$

(b) The value of the retirement savings account is found just like in part (a) only we use $P = 750$:

$$\$V = \$750\left[\frac{(1+0.066)^{45}-1}{0.066}\right]$$

$$\approx \$190,283.26.$$

Note that this is amount is exactly half of what was found in part (a) since the annual contributions are half of what they were in part (a). This illustrates the linear relationship between the value V of the retirement savings account and the annual contributions P.

(c) The value of the retirement savings account is found just like in part (a) only we use $P = 2250$:

$$\$V = \$2250\left[\frac{(1+0.066)^{45}-1}{0.066}\right]$$

$$\approx \$570,849.78.$$

Alternatively, one could simply add the values found in parts (a) and (b): $\$380,566.52 + \$190,283.26 = \$570,849.78.$

47. (a) We use the Retirement Savings Formula (with annual contributions),

$$\$V = \$P\left[\frac{(1+r)^{Y}-1}{r}\right], \text{ where the annual}$$

contribution (in dollars) made at the end of each year is $P = 1500$, the annual interest rate is $r = 0.066$, and the number of years of retirement savings is $Y = 40$. The value of the retirement savings account is then calculated to be

$$\$V = \$1500\left[\frac{(1+0.066)^{40}-1}{0.066}\right]$$

$$\approx \$270,250.07.$$

(b) Again, use the Retirement Savings Formula with $P = 1500$, $r = 0.066$, and $Y = 50$. Then,

$$\$V = \$1500\left[\frac{(1+0.066)^{50}-1}{0.066}\right]$$

$$\approx \$532,419.17.$$

(c) Use the Retirement Savings Formula with $P = 1500$, $r = 0.066$, and $Y = 52$. Then,

$$\$V = \$1500\left[\frac{(1+0.066)^{52}-1}{0.066}\right]$$

$$\approx \$608,116.71.$$

This exercise illustrates that the relationship between the value V of the retirement savings account and the number of years Y of annual contributions is not linear (as we know, this relationship is exponential).

49. (a) We use the Retirement Savings Formula (with monthly contributions),

$$\$V = \$M\frac{[1+(r/12)]^{T}-1}{(r/12)}, \text{ where the}$$

monthly contribution (in dollars) made at the end of each month is $M = 125$, the annual interest rate is $r = 0.066$, and the number of months of retirement savings is $T = 12(45) = 540$. The value of the retirement savings account is then

$$\$V = \$125\frac{[1+(0.066/12)]^{540}-1}{(0.066/12)}$$

$$\approx \$416,680.55.$$

Compare this value to that found in Exercise 45(a). The value of the account is much more even though the annual contributions were the same (just made at the end of each month rather than the end of each year).

(b) Use the Retirement Savings Formula (with monthly contributions) with $M = 62.50$, $r = 0.066$, and $T = 12(45) = 540$. Then,

$$\$V = \$62.50\frac{[1+(0.066/12)]^{540}-1}{(0.066/12)}$$

$$\approx \$208,340.28.$$

Compare this value to that found in Exercise 45(b).

(c) Use the Retirement Savings Formula (with monthly contributions) with $M = 187.50$, $r = 0.066$, and $T = 12(45) = 540$. Then,

$$\$V = \$187.50\frac{[1+(0.066/12)]^{540}-1}{(0.066/12)}$$

$$\approx \$625,020.83.$$

Compare this value to that found in Exercise 45(c). Alternatively, one could simply add the values found in parts (a) and (b): $\$416,680.55 + \$208,340.28 = \$625,020.83.$

51. (a) We use the Retirement Savings Formula (with monthly contributions),

$$\$V = \$M\frac{\left[1+(r/12)\right]^{T}-1}{(r/12)}, \text{ where the}$$

monthly contribution (in dollars) made at the end of each year is $M = 125$, the annual interest rate is $r = 0.066$, and the number of months of retirement savings is $T = 12(40) = 480$. The value of the retirement savings account is then calculated to be

$$\$V = \$125\frac{\left[1+(0.066/12)\right]^{480}-1}{(0.066/12)}$$

$$\approx \$293,459.20.$$

Compare this value to that found in Exercise 49(a).

(b) Use the Retirement Savings Formula with $M = 125$, $r = 0.066$, and $T = 12(50) = 600$. Then,

$$\$V = \$125\frac{\left[1+(0.066/12)\right]^{600}-1}{(0.066/12)}$$

$$\approx \$587,922.62.$$

(c) Use the Retirement Savings Formula with $M = 125$, $r = 0.066$, and $T = 12(52.5) = 630$. Then,

$$\$V = \$125\frac{\left[1+(0.066/12)\right]^{630}-1}{(0.066/12)}$$

$$\approx \$697,143.48.$$

This exercise illustrates that the relationship between the value V of the retirement savings account and the number of months T of monthly contributions is not linear (as we know, this relationship is exponential).

53. We use the Retirement Savings Formula (with monthly contributions),

$$\$V = \$M\frac{\left[1+(r/12)\right]^{T}-1}{(r/12)}, \text{ where the}$$

monthly contribution M (in dollars) made at the end of each year is unknown, the annual interest rate is $r = 0.066$, the number of months of retirement savings is $T = 12(40) = 480$, and the value of the retirement account at the end of this time is $V = 500,000$ (in dollars).

Since all the variables in this formula are known except for M, we solve

$$\$500,000 = \$M\frac{\left[1+(0.066/12)\right]^{480}-1}{(0.066/12)} \text{ to}$$

obtain a monthly contribution of

$$\$M = \frac{\$500,000(0.066/12)}{\left[1+(0.066/12)\right]^{480}-1} \approx \$212.98.$$

55. A table can help better organize the calculations needed in this automatic escalation retirement plan. Think of this as three plans (Plans A, B, and C). In Plan A, \$125 is contributed each month for 480 months (40 years). This leads to an ending value of

$$\$V = \$125\frac{\left[1+(0.066/12)\right]^{480}-1}{(0.066/12)} \approx \$293,459.20.$$

In Plan B, \$100 (\$225-\$125) is contributed each month for 240 months (20 years). This leads to an ending value of

$$\$V = \$100\frac{\left[1+(0.066/12)\right]^{240}-1}{(0.066/12)} \approx \$49,634.67.$$

In Plan C, \$175 (\$400-\$225) is contributed each month for 60 months (5 years). This leads to an ending value of

$$\$V = \$100\frac{\left[1+(0.066/12)\right]^{60}-1}{(0.066/12)} \approx \$12,399.90.$$

	Contribution	Years	Number of Months	Value
Plan A	\$125	1 to 40	480	\$293,459.20
Plan B	\$100	21 to 40	240	\$49,634.67
Plan C	\$175	35 to 40	60	\$12,399.90
Total				**\$355,493.77**

10.5 Consumer Debt

57. The amortization formula is given by

$$M = P\frac{p(1+p)^T}{(1+p)^T-1} \quad \text{where} \quad p = \frac{0.06}{12} = 0.005$$

is the monthly interest rate, M is the monthly payment, $P = \$14,500$ is the principal, and $T = 36$ is the number of monthly payments. So, the monthly payments are

$$M = \$14,500\frac{0.005(1+0.005)^{36}}{(1+0.005)^{36}-1} = \$441.12$$

59. The amortization formula is given by

$$M = P\frac{p(1+p)^T}{(1+p)^T-1} \quad \text{where} \quad p = \frac{0.036}{12} = 0.003$$

is the monthly interest rate, $M = \$400$ is the monthly payment, P is the principal (amount to be financed), and $T = 48$ is the number of monthly payments. Solving for the value of P to the nearest dollar gives

$$P = M\frac{(1+p)^T-1}{p(1+p)^T}$$

$$= \$400\frac{(1+0.003)^{48}-1}{0.003(1+0.003)^{48}}$$

$$\approx \$17,857.$$

61. (a) The amortization formula is given by

$$M = P\frac{p(1+p)^T}{(1+p)^T-1} \quad \text{where} \quad p = \frac{0.0575}{12} \text{ is}$$

the monthly interest rate, M is the monthly payment, $P = \$160,000$ is the principal, and $T = 30\times12 = 360$ is the number of monthly payments. So, the monthly payments are

$$M = \$160,000\frac{\frac{0.0575}{12}\left(1+\frac{0.0575}{12}\right)^{360}}{\left(1+\frac{0.0575}{12}\right)^{360}-1}$$

$$= \$933.72.$$

(b) To determine the amount of interest paid over the life of the loan, we calculate the amount paid over the course of 360 months and subtract the original principal of $160,000. This gives
$360\times\$933.72-\$160,000 = \$176,139.20.$

63. The amortization formula is given by

$$M = P\frac{p(1+p)^T}{(1+p)^T-1} \quad \text{where} \quad p = \frac{0.055}{12} \text{ is the}$$

monthly interest rate, $M = \$950$ is the monthly payment, P is the principal (amount financed), and $T = 240$ is the number of monthly payments. Solving for the value of P gives

$$P = M\frac{(1+p)^T-1}{p(1+p)^T}$$

$$= \$950\frac{\left(1+\frac{0.055}{12}\right)^{240}-1}{\frac{0.055}{12}\left(1+\frac{0.055}{12}\right)^{240}}$$

$$\approx \$138,104.$$

Adding the $25,000 down payment to the amount financed P will give the selling price of the house which is $138,104 + \$25,000 = \$163,104.$

JOGGING

65. (a) First, organize the information by periods of constant balance.

Period	Daily Balance	Days
6/19 – 6/20	$1200	2
6/21 – 6/29	1379.58	9
6/30 – 7/4	1419.58	5
7/5 – 7/14	1517.93	10
7/15 – 7/18	1017.93	4

The average daily balance is then easy to calculate. We simply add the daily balances and then divide by 30. The sum of the daily balances is 2(1200) + 9(1379.58) + 5(1419.58) + 10(1517.93) + 4(1017.93) = 41,165.14. Dividing by 30 gives an average daily balance of $1372.17.

(b) The periodic interest rate is
$\frac{30}{365}\times0.195 = 0.016027$. So, the finance charge is $1372.17\times0.016027 = \$21.99$.

(c) $1017.93+\$21.99 = \$1039.92$

67. Let m be the markup and c be the retailer's cost.

$$0.75(1+m)c = 1.5c$$
$$0.75 + 0.75m = 1.5$$
$$0.75m = 0.75$$
$$m = 1$$

The markup should be 100%.

69. (a) Another way to write $y\%$ is $y/100$ and another way to write $x\%$ is $x/100$. So, taking $y\%$ off an item is the same as multiplying the price of the item by $1-y/100$. Taking $x\%$ off an item is the same as multiplying the price by $1-x/100$. If the item originally cost \$$P$, then the result of these two discounts would be an item priced at $\$\left(1-\dfrac{x}{100}\right)\left[\left(1-\dfrac{y}{100}\right)P\right]$.

(b) $\$\left(1-\dfrac{y}{100}\right)\left[\left(1-\dfrac{x}{100}\right)P\right]$

(c) Multiplication is commutative. That is, it can be done in any order.

71. The loan (amount financed) from Bank A would be $P = \$90{,}000$ with a periodic (monthly) interest rate of $p = \dfrac{0.10}{12}$. We use the amortization formula $M = P\dfrac{p(1+p)^T}{(1+p)^T - 1}$ with $T = 360$ monthly payments of \$$M$. The loan from Bank A requires a monthly payment of

$$M = P\frac{p(1+p)^T}{(1+p)^T - 1}$$

$$= \$90{,}000\frac{\dfrac{0.1}{12}\left(1+\dfrac{0.1}{12}\right)^{360}}{\left(1+\dfrac{0.1}{12}\right)^{360} - 1}$$

$$\approx \$789.81$$

The loan from Bank B would be \$92,783.51 (leaving \$90,000 after paying the 3% loan fee). Using the same ideas as above, the monthly payment turns out to be \$780.17 for 360 months. Consequently, the loan from Bank B is \$9.64 a month less than the loan from Bank A, saving \$3470.40 over the 360 months. Bank B offers the better deal.

73. (a) First, we show that the increase in monthly payment that will pay the mortgage off in

21 years is indeed close to \$100 using the amortization formula. The monthly payment on a 30-year mortgage for \$100,000 with an APR of 6% would be

$$M = P\frac{p(1+p)^T}{(1+p)^T - 1}$$

$$= \$100{,}000\frac{\dfrac{0.06}{12}\left(1+\dfrac{0.06}{12}\right)^{360}}{\left(1+\dfrac{0.06}{12}\right)^{360} - 1}$$

$$\approx \$599.55 \text{ (call it \$600)}.$$

However, to pay off a \$100,000 in 21 years would require a payment of

$$M = P\frac{p(1+p)^T}{(1+p)^T - 1}$$

$$= \$100{,}000\frac{\dfrac{0.06}{12}\left(1+\dfrac{0.06}{12}\right)^{252}}{\left(1+\dfrac{0.06}{12}\right)^{252} - 1}.$$

$$\approx \$698.86 \text{ (call it \$700)}.$$

The interest paid over 30 years would be about $360 \times \$600 - \$100{,}000 = \$116{,}000$. The interest paid if the mortgage were paid off in 21 years would be $252 \times \$700 - \$100{,}000 = \$76{,}400$ which equates to a savings of \$40,000.

(b) To pay off a $P = \$100{,}000$ in 15 years ($T = 180$ months) with an APR of 6% would require a monthly payment of

$$M = P\frac{p(1+p)^T}{(1+p)^T - 1}$$

$$= \$100{,}000\frac{0.005(1.005)^{180}}{(1.005)^{180} - 1}$$

$$\approx \$843.86.$$

This is an increase of approximately $\$844 - \$600 = \$244$ each month (over a 30-year mortgage payment). The total interest paid on this 15-year mortgage is $180 \times \$843.86 - \$100{,}000 = \$51{,}894.80$ (call it \$51,895). The amount of interest saved in going from a 30- to a 15-year loan would therefore be about $\$116{,}000 - \$51{,}895 = \$64{,}105$.

(c) To pay off a $P = \$100{,}000$ mortgage in t ($t < 30$) years at an APR of 6% would require a payment of

$$M = P\frac{p(1+p)^T}{(1+p)^T - 1}$$

$$= \$100{,}000\frac{0.005(1.005)^{12t}}{(1.005)^{12t} - 1}$$

$$= \$500\frac{(1.005)^{12t}}{(1.005)^{12t} - 1}$$

This is an increase of

$$\$500\frac{(1.005)^{12t}}{(1.005)^{12t} - 1} - \$600 \text{ each month.}$$

75. (a) Let $V = P\left[\dfrac{(1+r)^Y - 1}{r}\right]$ be the value of the retirement savings account with annual contribution P. Then the value of a retirement savings account with annual contribution cP is

$$cP\left[\frac{(1+r)^Y - 1}{r}\right] = c\left(P\left[\frac{(1+r)^Y - 1}{r}\right]\right) = cV.$$

(b) Let $V = P\left[\dfrac{(1+r)^Y - 1}{r}\right]$ be the value of the retirement savings account with annual contribution P, and let $W = Q\left[\dfrac{(1+r)^Y - 1}{r}\right]$ be the value of the retirement savings account with annual contribution Q. Then the value of a retirement savings account with annual contribution $(P+Q)$ is $(P+Q)\left[\dfrac{(1+r)^Y - 1}{r}\right]$

$$= P\left[\frac{(1+r)^Y - 1}{r}\right] + Q\left[\frac{(1+r)^Y - 1}{r}\right] = V + W.$$

Chapter 11

WALKING

11.2 Reflections

1. (a)

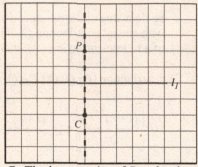

(b) *F*. See the figure below.

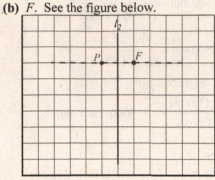

C. The image point of *P* under the reflection with axis l_1 is found by drawing a line through *P* perpendicular to l_1 and finding the point on this line on the opposite side of l_1 which is the same distance from l_1 as the point *P*. This point is *C*.

(c) *E*. See the figure below.

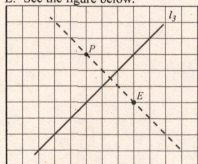

(d) *B*. See the figure below.

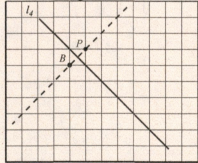

3. (a)

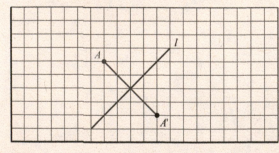

Remember that the axis of reflection needs to be perpendicular to the line segment connecting *A* and *A'*. That is, *l* is the perpendicular bisector of *AA'*.

(b) The image of *A'* under the reflection is *A*. (*A* reflects to *A'* which reflects about *l* back to *A*)

5. (a)

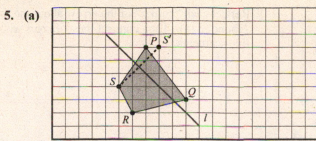

Remember that the axis of reflection needs to be perpendicular to the line segment connecting S and S'. That is, l is the perpendicular bisector of SS'.

(b)

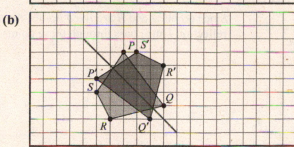

Reflect each point P, Q and R about the line l just as you did for S in (a). Then connect the dots to form the quadrilateral $P'Q'R'S'$.

7. (a)
(b)

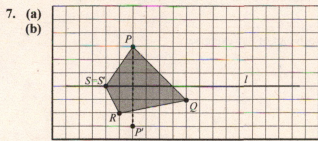

The axis of reflection needs to be perpendicular to the line segment connecting P and P'. That is, l is the perpendicular bisector of PP'.

Since S is on the axis of reflection l, we see that S', the image of S under the reflection, is at the same position as S.

(c)

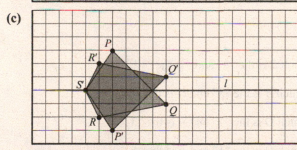

Reflect each point P, Q, R, and S about the line l found in (a).

(d)

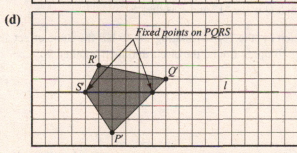

Fixed points on PQRS

Any point that is on both the axis of reflection l and the quadrilateral $PQRS$ is a fixed point of the reflection.

9. (a)

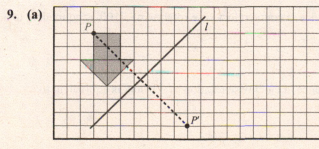

The axis of reflection is perpendicular to the line segment connecting P and P'. That is, l is the perpendicular bisector of PP'.

(b)

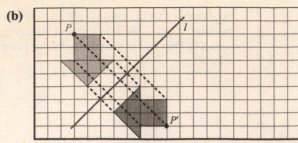

Reflect each "corner" of the arrow about the line *l* found in (a). Then connect these images with line segments.

11.

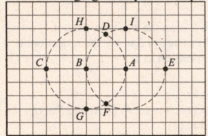

Since points *A* and *B* are fixed points, the axis of reflection *l* must pass through these points.

11.3 Rotations

13. The following figure may be of help in solving each part of this exercise.

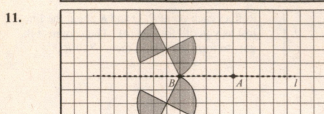

(a) *I*. Think of *A* as the center of a clock in which *B* is the "9" and *I* is the "12."

(b) *G*. Think of *B* as the center of a clock in which *A* is the "3" and *G* is the "6."

(c) *A*. Think of *B* as the center of a clock in which *D* is the "1" and *A* is the "3."

(d) *F*. Think of *B* as the center of a clock in which *D* is the "1" and *F* is the "5."

(e) *E*. Think of *A* as the center of a clock in which *I* is the "12." Rotating $3690°$ has the same effect as rotating $10 \times 360° + 90°$. That is, as rotating 10 times around the circle and then another $90°$.

15. (a) $2(360°) - 710° = 10°$

(b) $710° - 360° = 350°$

(c) $360°\overline{)7100°}$ with a quotient of 19 and a remainder of $260°$. Hence, a counterclockwise rotation of $7100°$ is equivalent to a clockwise rotation of $360° - 260° = 100°$.

(d) $360°\overline{)71,000°}$ with a quotient of 197 and a remainder of $80°$. Hence, a clockwise rotation of $71,000°$ is equivalent to a clockwise rotation of $80°$.

17. (a)
(b)

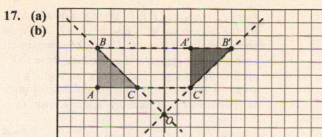

Since *BB'* and *CC'* are parallel, the intersection of *BC* and *B'C'* locates the rotocenter *O*. This is a 90° clockwise rotation.

[Note: The vertical line is the perpendicular bisector of both *BB'* and *CC'*.]

19. (a)

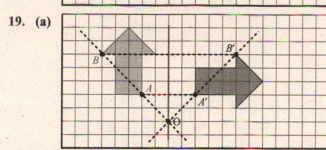

Since *AA'* and *BB'* are parallel, the intersection of *AB* and *A'B'* locates the rotocenter *O*. This is a 90° clockwise rotation.

[Note: The vertical line is the perpendicular bisector of both *AA'* and *BB'*.]

(b)

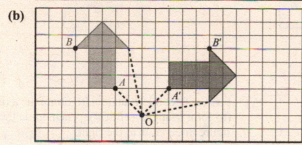

Rotate each "corner" of the arrow clockwise 90°.

21.

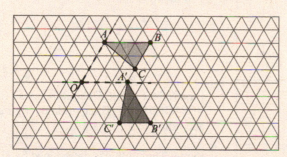

The equilateral triangles that make up the grid have interior angles that each measure 60°. So the triangle *ABC* is being rotated 60° clockwise.

11.4 Translations

23. (a) *C*. Vector v_1 translates a point 4 units to the right so the image of *P* is *C*.

(b) *C*. Vector v_2 translates a point 4 units to the right so the image of *P* is *C*.

(c) *A*. Vector v_3 translates a point up 2 units and right 1 unit so the image of *P* is *A*.

(d) *D*. Vector v_4 translates a point down 2 units and left 1 unit so the image of *P* is *D*.

25. (a)
(b)
(c)

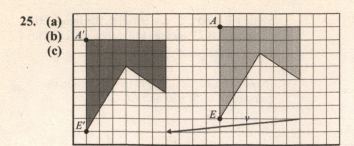

Point E' is located left 10 units and down 1 unit from E. So, point A' is also located left 10 units and down 1 unit from A. Doing the same for all corners of the region and connecting them will produce the image. The translation vector v is an arrow pointing left 10 units and down 1 unit. We can illustrate it from any starting position.

27.

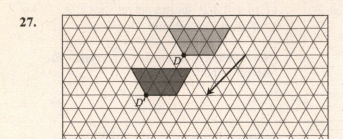

11.5 Glide Reflections

29.

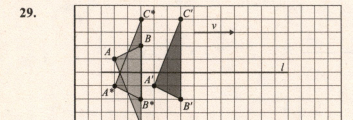

First, reflect the triangle ABC about the axis l (to form triangle $A*B*C*$). Then, glide the figure three units to the right.

31.

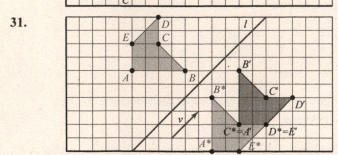

First reflect $ABCDE$ about the axis l (to form $A*B*C*D*E*$). Then, glide the figure according to v.

[Note: The glide v is determined by $D*D'$ which is easily seen after the reflection.]

33. (a)
(b)

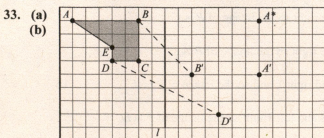

The midpoints of line segments BB' and DD' determine the axis of reflection for this glide reflection.

To determine the location of A', reflect A across l to a point $A*$. Then, glide $A*$ down 4 units to A'.

[Note: B reflects to a point 4 units above B'.]

(c)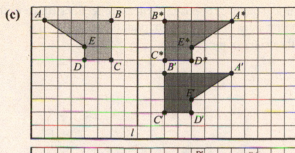

Reflect the figure *ABCDE* about the axis of reflection *l* (to form *A*B*C*D*E**). Then, glide the figure four units down.

35.

The midpoints of line segments *PP'* and *QQ'* usually determine the axis of reflection for a glide reflection. However, in this case, the midpoints are the same point. So the line the *PQ* is drawn and the line perpendicular to *PQ* that passes through the midpoints is the axis of reflection *l*. We then reflect the figure about this axis *l* and glide the result 4 units to the left (so that *P** lands on *P'*).

37.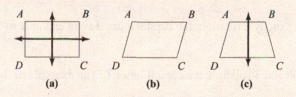

The midpoints of line segments *CC'* and *DD'* determine the axis of reflection *l* for this glide reflection. First, reflect the figure about this 60-degree axis (the result is the figure labeled with *C** and *D**). Then, glide down and to the right.

11.6 Symmetries and Symmetry Types

39. The following figure gives the axes of reflection for each figure.

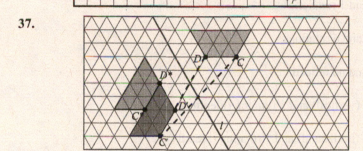

(a)   (b)   (c)

(a) Reflection with axis going through the midpoints of *AB* and *DC*; reflection with axis going through the midpoints of *AD* and *BC*; rotations of 180° and 360° with rotocenter the center of the rectangle.

(b) No Reflections. Rotations of 180° and 360° with rotocenter the center of the parallelogram.

(c) Reflection with axis going through the midpoints of *AB* and *DC*; rotation of 360° with rotocenter the center of the trapezoid.

41. (a) Reflections (three of them) with axis going through pairs of opposite vertices; reflections (three of them) with axis going through the midpoints of opposite sides of the hexagon; rotations of 60°, 120°, 180°, 240°, 300°, 360° with rotocenter the center of the hexagon.

(b) No reflections; rotations of 72°, 144°, 216°, 288°, 360° with rotocenter the center of the star.

43. (a) D_2; the figure has exactly 2 reflections and 2 rotations.

(b) Z_2; the figure has no reflections and exactly 2 rotations.

(c) D_1; the figure has exactly 1 reflection and 1 rotation.

45. (a) D_6; the figure has exactly 6 reflections and 6 rotations.

(b) Z_5; the figure has exactly 5 rotations (and no reflections).

47. (a) D_1; the letter **A** has exactly 1 reflection (vertical) and 1 rotation (identity).

(b) D_1; the letter **D** has exactly 1 reflection (horizontal) and 1 rotation (identity).

(c) Z_1; the letter **L** has no reflection and exactly 1 rotation (identity).

(d) Z_2; the letter **Z** has no reflection and exactly 2 rotations (identity and 180°).

(e) D_2; the letter **H** has exactly 2 reflections and 2 rotations (identity and 180°).

(f) Z_2; the letter **N** has no reflection and exactly 2 rotations (identity and 180°).

49. Answers will vary.

(a) Since symmetry type Z_1 has no reflections and exactly 1 rotation, the capital letter **J** is an example of this symmetry type.

(b) Since symmetry type D_1 has exactly 1 reflection and 1 rotation, the capital letter **T** is an example of this symmetry type.

(c) Since symmetry type Z_2 has no reflection and exactly 2 rotations, the capital letter **Z** is an example of this symmetry type.

(d) Since symmetry type D_2 has exactly 2 reflections and 2 rotations, the capital letter **I** is an example of this symmetry type.

11.7 Patterns

51. **(a)** $m1$; the border pattern has translation symmetry and vertical reflection but does not have horizontal reflection, half-turn rotation, or glide reflection.

    (b) $1m$; the border pattern has translation symmetry and horizontal reflection but does not have vertical reflection, half-turn, or glide reflection.

    (c) 12; the border pattern has translation symmetry and half-turn rotation but does not have horizontal reflection, vertical reflection, or glide reflection.

    (d) 11; the border pattern has translation symmetry but does not have horizontal reflection, vertical reflection, half-turn rotation, or glide reflection.

53. **(a)** $m1$; the border pattern has translation symmetry and vertical reflection but does not have horizontal reflection, glide reflection, or half-turn rotation.

    (b) 12; the border pattern has translation symmetry and half-turn rotation but does not have horizontal reflection, vertical reflection, or glide reflection.

    (c) $1g$; the border pattern has translation symmetry and glide reflection but does not have horizontal reflection, vertical reflection, or half-turn rotation.

    (d) mg; the border pattern has translation symmetry, vertical reflection, half-turn rotation, and glide reflection, but does not have horizontal reflection.

55. 12; see, for example, exercise 51(c).

57. mm; consider as an example ...$+ + + + + +$ .... Since the pattern has horizontal reflection symmetry, a border pattern of type $\_\_m$ is guaranteed. Since the motif has symmetry type D_4, we know that we must have vertical reflection symmetry (and two diagonal symmetries) too. That rules out $m1$. The axes of reflection for the reflection symmetries are 90-degree rotations of each other. So, the new type is mm.

JOGGING

59. **(a)** Long division gives $360 \overline{)\,7080\,}$ $\overset{\text{19 R 240}}{}$. That is, $7080°$ is 19 full revolutions and $240°$ more. This means that 19 full hours have passed.

    (b) Using the remainder found in (a), the minute hand points to the number on the clock that is located $240°$ (clockwise) from the 9. That is, it is two-thirds of the way around the face from 9 which is a full 8 of the 12 markings on the face of the clock. Rotating 8 of these markings from the 9 leaves the minute hand on the 5.

61. **(a)** rotation; the two rigid motions that are proper are rotations and translations. Of these, only a rotation can have exactly one fixed point (the identity translation has infinitely many fixed points).

    (b) identity motion; the two rigid motions that are proper are rotations and translations. Of these, only the identity motion has infinitely many fixed points.

    (c) reflection; the two rigid motions that are improper are reflections and glide reflections. But glide reflections do not have fixed points while reflections can have an infinite number of them.

    (d) glide reflection; the two rigid motions that are improper are reflections and glide reflections. Of these, only the glide reflection has no fixed points.

63. (a) *C.* The reflection of *P* about *l* is the point *B.* The image of *B* under the rotation is *C.*

(b) *P.* The image of *P* under the rotation is *B.* The reflection of *B* about *l* is *P.*

(c) *D.* The reflection of *P* about *l* is the point *B.* The rotation of *B* is *D.*

(d) *D.* The rotation of *P* is the point *C.* The reflection of *C* about line *l* is *D.* [Note that (a), (b), (c) and (d) suggest that the order in which reflections occur *sometimes* makes a difference.]

65. The combination of two improper rigid motions is a proper rigid motion. Since *C* is a fixed point, the rigid motion must be a rotation with rotocenter *C.*

67. (a) The result of applying the reflection with axis l_1, followed by the reflection with axis l_2, is a clockwise rotation with center *C* and angle of rotation $\lambda + \lambda + \beta + \beta = 2(\lambda + \beta) = 2\alpha$. One example is shown in the figure.

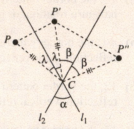

(b) The result of applying the reflection with axis l_2, followed by the reflection with axis l_1, is a *counter-clockwise* rotation with center *C* and angle of rotation 2α.

69. (a)

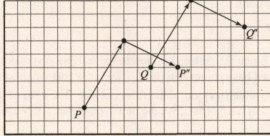

P″ is the image of *P* under the product of $\mathcal{M}$ and $\mathcal{N}$.

Q″ is the image of *Q* under the product of $\mathcal{M}$ and $\mathcal{N}$.

(b)

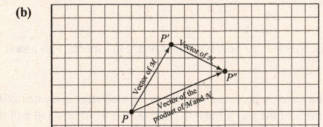

The vector of the translation representing the product of $\mathcal{M}$ and $\mathcal{N}$ is described by the arrow from *P* to *P″*.

71. Rotations and translations are proper rigid motions, and hence preserve clockwise-counterclockwise orientations. The given motion is an improper rigid motion (it reverses the clockwise-counterclockwise orientation). If the rigid motion was a reflection, then *PP′, RR′,* and *QQ′* would all be perpendicular to the axis of reflection and hence would all be parallel. It must be a glide reflection (the only rigid motion left).

RUNNING

73.

	r_1	r_2	r_3	R_1	R_2	I
r_1	I	R_1	R_2	r_2	r_3	r_1
r_2	R_2	I	R_1	r_3	r_1	r_2
r_3	R_1	R_2	I	r_1	r_2	r_3
R_1	r_3	r_1	r_2	R_2	I	R_1
R_2	r_2	r_3	r_1	I	R_1	R_2
I	r_1	r_2	r_3	R_1	R_2	I

75. *pm*; The identity is the only rotation symmetry. There is a reflection (vertical). However, there is not a glide reflection with an axis that is not a reflection axis.

77. *cm*; The identity is the only rotation symmetry. There is a reflection (vertical). There are also glide reflections that do not have the same axis as the vertical reflection axes. These vertical axes are ¼ and ¾ of the distance between two horizontal hearts.

79. *p4m*; The pattern has 4-fold rotational symmetry. There are reflections. In fact, there are reflections that intersect at 45° angles.

Chapter 12

WALKING

12.1 The Koch Snowflake and Self-Similarity

1.

	M	l	P
Start	3	81 cm	243 cm
Step 1	12	27 cm	324 cm
Step 2	48	9 cm	432 cm
Step 3	192	3 cm	576 cm
Step 4	768	1 cm	768 cm
Step 5	3072	1/3 cm	1024 cm

At each step, the value of the number of sides M is multiplied by 4 and the value of each side length l is multiplied by 1/3. The perimeter P at each step is simply the product of M and l.

3.

	R	S	T	Q
Start	0	0	0	81
Step 1	3	9	27	108
Step 2	12	1	12	120
Step 3	48	1/9	16/3	376/3
Step 4	192	1/81	64/27	3448/27
Step 5	768	1/729	256/243	31228/243

In the first step, 3 triangles are added. At each step thereafter, R (the number of triangles added) is multiplied by 4. The area S of each triangle added at a given step is 1/9 of the area added during the previous step. The total new area T added at a given step is then the product of R and S. The area of the "snowflake" obtained at a given step (which we call Q) is the sum of T and the area Q of the snowflake in the previous step.

For example, at step 1 a total of $R = 3$ triangles are added each having area $S = 9$ (S is 1/9 the area of the original triangle which was 81). So, the value of Q at step 1 is $81 + 3 \times 9 = 108$.

At step 2, $R = 3 \times 4 = 12$ triangles are added each having area $S = 1$ (this can be seen as either 1/9 of the area added in step 1 or $1/9^2$ the area of the original triangle). The value of Q at step 2 is then $Q = 108 + 12 \times 1 = 120$.

At step 3, $R = 3 \times 4^2 = 48$ triangles are added each having area $S = 1/9$ (again, this can be seen as 1/9 of the previous value of S). The value of Q at step 3 is then $Q = 120 + 48 \times (1/9) = 376/3$.

5.

	M	*l*	*P*
Start	4	81 cm	324 cm
Step 1	20	27 cm	540 cm
Step 2	100	9 cm	900 cm
Step 3	500	3 cm	1500 cm
Step 4	2500	1 cm	2500 cm

At each step, the value of the number of sides *M* is multiplied by 5 and the value of each side length *l* is multiplied by 1/3 since each side of the object is replaced by 5 sides having 1/3 the length. The perimeter *P* at each step is simply the product of *M* and *l*.

7.

	R	*S*	*T*	*Q*
Start	0	0	0	81
Step 1	4	9	36	117
Step 2	20	1	20	137
Step 3	100	1/9	100/9	1333/9
Step 4	500	1/81	500/81	12497/81

In the step 1, 4 squares are added. At each step thereafter, *R* (the number of squares added) is multiplied by 5. The area *S* of each square added at a given step is 1/9 of the area of each square added during the previous step. The total new area *T* added at a given step is then the product of *R* and *S*. The area of the shape obtained at a given step (which we call *Q*) is the sum of *T* and the area *Q* of the shape in the previous step.

At step 1, for example, a total of $R = 4$ squares are added each having area $S = 9$ (*S* is 1/9 the area of the original square which was 81). So, the value of *Q* at step 1 is $81 + 4 \times 9 = 117$.
At step 2, $R = 4 \times 5 = 20$ squares are added each having area $S = 1$ (this can be seen as 1/9 of the area added in step 1). The value of *Q* at step 2 is then $Q = 117 + 20 \times 1 = 137$.

At step 3, $R = 4 \times 5^2 = 100$ squares are added each having area $S = 1/9$ (again, this can be seen as 1/9 of the previous value of *S*). The value of *Q* at step 3 is then $Q = 137 + 100 \times (1/9) = 1333/9$.

9.

	M	*l*	*P*
Start	3	81 cm	243 cm
Step 1	12	27 cm	324 cm
Step 2	48	9 cm	432 cm
Step 3	192	3 cm	576 cm
Step 4	768	1 cm	768 cm
Step 5	3072	1/3 cm	1024 cm

At each step, the value of the number of sides *M* is multiplied by 4 (each side becomes four) and the value of each side length *l* is multiplied by 1/3 (each side length is divided into three equal lengths). The perimeter *P* at each step is simply the product of *M* and *l*.

11.

	R	**S**	**T**	**Q**
Start	0	0	0	81
Step 1	3	9	27	54
Step 2	12	1	12	42
Step 3	48	1/9	48/9	110/3
Step 4	192	1/81	192/81	926/27
Step 5	768	1/729	768/729	8078/243

In the first step, 3 triangles are subtracted. At each step thereafter, the number of triangles subtracted (which we call R) is multiplied by 4. The area S of each triangle subtracted at a given step is 1/9 of the area subtracted during the previous step. The total new area T subtracted at a given step is then the product of R and S. The area of the shape obtained at a given step (which we call Q) is the difference of the area of the shape in the previous step and T. [Note: This is similar to exercise 3 except subtraction is used rather than addition.]

For example, at step 1 a total of $R = 3$ triangles are subtracted each having area $S = 9$ (S is 1/9 the area of the original triangle which was 81). So, the value of Q at step 1 is $81 - 3 \times 9 = 54$.

At step 2, $R = 3 \times 4 = 12$ triangles are subtracted each having area $S = 1$. The value of Q at step 2 is then $Q = 54 - 12 \times 1 = 42$.

At step 3, $R = 3 \times 4^2 = 48$ triangles are subtracted each having area $S = 1/9$ (again, this can be seen as 1/9 of the previous value of S). The value of Q at step 3 is then $Q = 42 - 48 \times (1/9) = 110/3$.

13. (a)

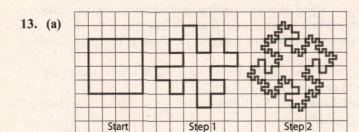

Start   Step 1   Step 2

(b) The seed square has perimeter of $4 \times 16 = 64$. By replacing each segment with the "sawtooth" version, the length is doubled. So, the perimeter of the entire figure obtained in step 1 is $2 \times 64 = 128$.

(c) As in (b), replacing each segment with the "sawtooth" version doubles the length. So, since the perimeter in step 1 is 128, the perimeter of the figure obtained in step 2 is $2 \times 128 = 256$.

(d) The perimeter doubles at each step of the construction.

15. (a) 256; the area at step 1 is the same as the area of the seed (what the rule "giveth," the rule "taketh" away).

(b) 256; the area at step 2 is the same as the area after step 1.

(c) At each step, the area added is the same as the area subtracted.

12.2 The Sierpinski Gasket and the Chaos Game

17.

	R	S	T	Q
Start	0	0	0	64
Step 1	1	16	16	48
Step 2	3	4	12	36
Step 3	9	1	9	27
Step 4	27	1/4	27/4	81/4
Step 5	81	1/16	81/16	243/16

In the step 1, one triangle is subtracted. At each step thereafter, the number of triangles subtracted (*R*) is multiplied by 3. Starting with step 2, the area *S* of each triangle subtracted at a given step is 1/4 of the area of each triangle subtracted during the previous step (in step 1, the area of the triangle that is subtracted is 1/4). The total new area *T* subtracted at a given step is the product of *R* and *S* at that step. The area of the shape obtained at a given step (*Q*) is the difference between the area of the shape in the previous step and *T*.

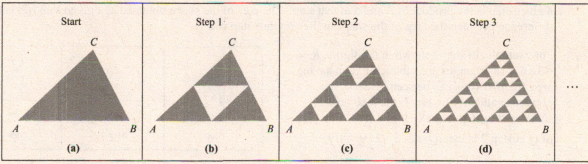

For example, at step 3 shown in (d) in the figure, $R = 1 \times 3^2 = 9$ triangles are subtracted each having area $S = 1$ (this can be seen as 1/4 of the previous value of *S*). So, *T*, the total area subtracted at step 3 is $9 \times 1 = 9$. The value of *Q* at step 3 is then $Q = 36 - 9 = 27$.

19.

	U	V	W
Start	1	8 cm	8 cm
Step 1	3	4 cm	12 cm
Step 2	9	2 cm	18 cm
Step 3	27	1 cm	27 cm
Step 4	81	1/2 cm	81/2 cm
Step 5	243	1/4 cm	243/4 cm

At each step, the value of the number of solid triangles *U* is tripled and the value of *V* (the perimeter of each solid triangle) is halved (since each side length is halved). The length of the boundary of the "gasket" *W* at each step is simply the product of *U* and *V*.

21.

	R	S	T	Q
Start	0	0	0	1
Step 1	3	1/9	1/3	2/3
Step 2	18	1/81	2/9	4/9
Step 3	108	1/729	4/27	8/27
Step 4	648	1/6561	8/81	16/81
Step N	$3 \cdot 6^{N-1}$	$1/9^N$	$3 \cdot 6^{N-1}/9^N$	$(2/3)^N$

In step 1, three triangles are subtracted. In each step thereafter, the number of triangles subtracted (R) is multiplied by 6 (since six solid triangles remain for every three that are subtracted). Starting in step 2, the area S of each triangle subtracted at a given step is 1/9 of the area of each triangle subtracted during the previous step (in step 1, the area of each triangle that is subtracted is 1/9). The total new area T subtracted at a given step is the product of R and S at that step. The area of the shape obtained at a given step (Q) is the difference between the area of the shape in the previous step and T.

For example, in step 2 shown in the figure, R $= 3 \times 6 = 18$ triangles are subtracted each having area $S = 1/81$ (this can be seen as 1/9 of the previous value of S). So, T, the total area subtracted in step 2 is $18 \times 1/81 = 2/9$. The value of Q at step 2 is then $Q = 2/3 - 2/9 = 4/9$.

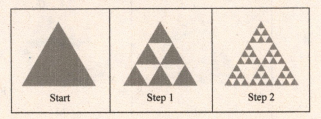

23.

	U	V	W
Start	1	9 cm	9 cm
Step 1	6	3 cm	18 cm
Step 2	36	1 cm	36 cm
Step 3	216	1/3 cm	72 cm
Step 4	1296	1/9 cm	144 cm
Step N	6^N	$1/3^{N-2}$ cm	$9 \cdot 2^N$ cm

At each step, the value of the number of solid triangles U is multiplied by 6 and the value of V (the perimeter of each solid triangle) is 1/3 of that in the previous step (since each side length is also 1/3 of that in the previous step). The length of the boundary of the "ternary gasket" W at each step is simply the product of U and V.

25. (a) 5/9. The solid square is divided into 9 equal subsquares and 4 of these (along the sides) are removed.

(b) $(5/9)^2$. The area of each of the subsquares found in step 2 is 5/9 of its area in step 1.

(c) $(5/9)^N$. The area of each of the subsquares found in step N is 5/9 of its area in step N-1. Since the area in step 1 is 5/9, the area obtained in step N is found through repeated multiplication (N-1 times) by 5/9.

27. The coordinates of each point are:

P_1 : (32, 0);

P_2 : (16, 0), the midpoint of A and P_1;

P_3 : (8, 16), the midpoint of C and P_2;

P_4 : (20, 8), the midpoint of B and P_3;

P_5 : (10, 20), the midpoint of C and P_4;

P_6 : (5, 26), the midpoint of C and P_5.

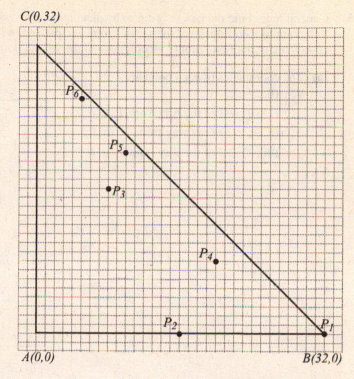

29.

Roll	Point	Coordinates
3	P_1	(32,0)
1	P_2	(16,0)
2	P_3	(8,0)
3	P_4	(20,0)
5	P_5	(10,16)
5	P_6	(5,24)

$P_3 = (8, 0)$ is the midpoint of A and P_2; $P_4 = (20,0)$ is the midpoint of B and P_3; $P_5 = (10, 16)$ is the midpoint of C and P_4; $P_6 = (5, 24)$, the midpoint of C and P_5.

31. General note: Find the new coordinate by picking the new *x*-value to be 2/3 of way from the *x*-coordinate of first point to *x*-coordinate of the second point and picking the new *y*-value to be 2/3 of the way from *y*-coordinate of the first point to *y*-coordinate of the second point.

(a) The coordinates of each point are:

P_1 : (0, 27);

P_2 : (18, 9), 2/3 of the way from P_1 to *B*;

P_3 : (6, 3), 2/3 of the way from P_2 to *A*;

P_4 : (20, 1), 2/3 of the way from P_3 to *B*.

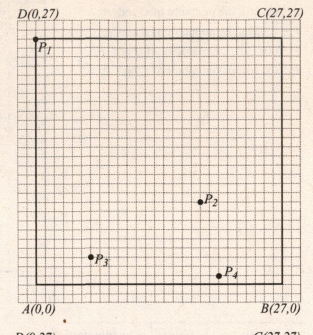

(b) The coordinates of each point are:

P_1 : (27, 27);

P_2 : (27, 9), 2/3 of the way from P_1 to *B*;

P_3 : (9, 3), 2/3 of the way from P_2 to *A*;

P_4 : (21, 1), 2/3 of the way from P_3 to *B*.

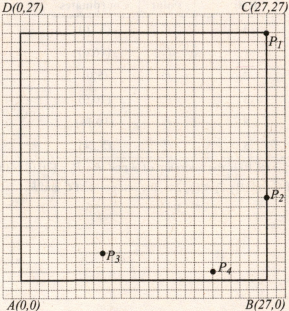

(c) The coordinates of each point are:

P_1 : (27, 27);

P_2 : (27, 27);

P_3 : (9, 9), 2/3 of the way from P_2 to A;

P_4 : (3, 3), 2/3 of the way from P_3 to A.

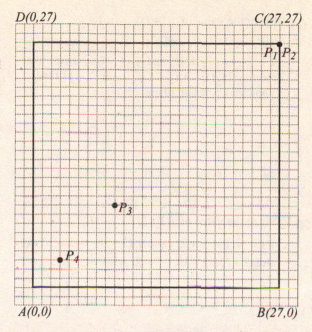

33. **(a)** 4,2,1,2; $P_1 = (0,27)$ corresponds to a first roll of 4. Then, since $P_2 = (18,9)$ is 2/3 of the way from P_1 to B, we see that the second roll is 2. Since $P_3 = (6,3)$ is 2/3 of the way from P_2 to A, we see that the third roll is 1. Lastly, since $P_4 = (20,1)$ is 2/3 of the way from P_3 to B, the fourth and final roll is a 2.

(b) 3,1,1,3; $P_1 = (27,27)$ corresponds to a first roll of 3. Then, since $P_2 = (9,9)$ is 2/3 of the way from P_1 to A, we see that the second roll is 1. Since $P_3 = (3,3)$ is 2/3 of the way from P_2 to A, we see that the third roll is 1. Lastly, since $P_4 = (19,19)$ is 2/3 of the way from P_3 to C, the fourth and final roll is a 3.

(c) 1,3,4,2; $P_1 = (0,0)$ corresponds to a first roll of 1. Then, since $P_2 = (18,18)$ is 2/3 of the way from P_1 to C, we see that the second roll is 3. Since $P_3 = (6,24)$ is 2/3 of the way from P_2 to D, we see that the third roll is 4. Lastly, since $P_4 = (20,8)$ is 2/3 of the way from P_3 to B, the fourth and final roll is a 2.

12.4 The Mandelbrot Set

35. **(a)** $(-i)^2 + (-i) = i^2 - i = -1 - i$

(b) $(-1-i)^2 + (-i) = (-1)^2 + 2i + i^2 + (-i) = 1 + 2i - 1 - i = i$

(c) $i^2 + (-i) = -1 - i$

37. **(a)** $(-0.25 + 0.25i)^2 + (-0.25 + 0.25i) = 0.0625 - 0.125i - 0.0625 + (-0.25 + 0.25i) = -0.25 + 0.125i$

(b) $(-0.25 - 0.25i)^2 + (-0.25 - 0.25i) = 0.0625 + 0.125i - 0.0625 + (-0.25 - 0.25i) = -0.25 - 0.125i$

39. (a) Since the value of $i(1+i) = -1+i$,
$i^2(1+i) = -1-i$, and $i^3(1+i) = 1-i$ we plot
$1+i$, $-1+i$, $-1-i$, and $1-i$. This is just like
plotting $(1,1)$, $(-1,1)$, $(-1,-1)$, and $(1,-1)$ in the
standard Cartesian plane.

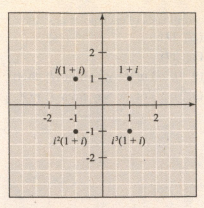

(b) Since the value of $i(3-2i) = 2+3i$,
$i^2(3-2i) = -3+2i$, and $i^3(3-2i) = -2-3i$ we
plot $3-2i$, $2+3i$, $-3+2i$, and $-2-3i$.

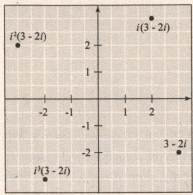

(c) The effect of multiplying each point in (a) and (b) by i is a 90-degree counterclockwise rotation.

41. (a) $s_1 = (-2)^2 + (-2) = 4 - 2 = 2$; $s_2 = (2)^2 + (-2) = 4 - 2 = 2$;
$s_3 = (2)^2 + (-2) = 4 - 2 = 2$; $s_4 = (2)^2 + (-2) = 4 - 2 = 2$.

(b) The sequence is attracted to 2 (in fact, each term equals 2), so $s_{100} = 2$.

(c) Each number in the sequence is 2. In this case, the sequence could be considered periodic (with period 1) or attracted to 2 (in the sense that a sequence of 2's could be considered getting closer and closer to the attractor 2).

43. (a) $s_1 = (-0.5)^2 + (-0.5) = -0.25$; $s_2 = (-0.25)^2 + (-0.5) = -0.4375$; $s_3 = (-0.4375)^2 + (-0.5) \approx -0.3086$;
$s_4 = (-0.3086)^2 + (-0.5) \approx -0.4048$; $s_5 = (-0.4048)^2 + (-0.5) \approx -0.3361$.

(b) $s_{N+1} = (-0.366)^2 + (-0.5) \approx -0.366$

(c) From part (b), we see that $s_N = s_{N+1}$, so the sequence must be attracted to -0.3660 (to 4 decimal places).

45. (a) For the first three values, see Exercise 35. $s_1 = (-i)^2 + (-i) = i^2 - i = -1 - i$;
$s_2 = (-1-i)^2 - i = 1 + 2i + i^2 - i = i$; $s_3 = i^2 - i = -1 - i$; $s_4 = (-1-i)^2 - i = 1 + 2i + i^2 - i = i$;
$s_5 = i^2 - i = -1 - i$.

(b) Periodic; the odd terms are $-1-i$ and the even terms are i.

JOGGING

47. (a) As in Exercise 18, we see that the value of R at step N is 3^{N-1}, the value of S at step N is $A/4^N$. So, the

value of T at step N of the construction is $3^{N-1} \cdot \dfrac{A}{4^N} = \dfrac{A}{4} \cdot \left(\dfrac{3}{4}\right)^{N-1}$. It follows that Q, the area of the

"gasket" obtained at step N of the construction is

$$A - \frac{A}{4} - \frac{A}{4} \cdot \left(\frac{3}{4}\right)^1 - \frac{A}{4} \cdot \left(\frac{3}{4}\right)^2 - \cdots - \frac{A}{4} \cdot \left(\frac{3}{4}\right)^{N-1} = A - \left[\frac{A}{4} + \frac{A}{4} \cdot \left(\frac{3}{4}\right)^1 + \frac{A}{4} \cdot \left(\frac{3}{4}\right)^2 + \cdots + \frac{A}{4} \cdot \left(\frac{3}{4}\right)^{N-1}\right]$$

$$= A - \frac{A}{4} \left[\frac{1 - \left(\frac{3}{4}\right)^N}{1 - \frac{3}{4}}\right] = A - A + A \cdot \left(\frac{3}{4}\right)^N = A \cdot \left(\frac{3}{4}\right)^N.$$

(b) The area of the Sierpinski gasket is smaller than the area of the gasket formed during any step of construction. That is, if the area of the original triangle is 1, then the area of the Sierpinski gasket is less

than $\left(\dfrac{3}{4}\right)^N$ for every positive value of N. Since $0 < 3/4 < 1$, the value of $\left(\dfrac{3}{4}\right)^N$ can be made smaller

than any positive quantity for a large enough choice of N. It follows that the area of the Sierpinski gasket can also be made smaller than any positive quantity.

49. (a)

	C	U	V
Start	0	0	1
Step 1	7	1/27	20/27
Step 2	20×7	$(1/27)^2$	$(20/27)^2$
Step 3	$20^2 \times 7$	$(1/27)^3$	$(20/27)^3$
Step 4	$20^3 \times 7$	$(1/27)^4$	$(20/27)^4$
Step N	$20^{N-1} \times 7$	$(1/27)^N$	$(20/27)^N$

At the first step, a cube is removed from each of the six faces and from the center for a total of $C = 7$ cubes removed. At the second step, each of the 20 remaining cubes has 7 cubes removed (i.e, $C = 20 \times 7$ cubes are removed). In step 3, there are 20^2 remaining cubes each of which has 7 cubes removed (i.e. $C = 20^2 \times 7$ cubes are removed). Etc.

Since step 1 can be thought of as $3 \times 3 \times 3 = 27$ equally sized cubes, the volume U of the middle cube removed is 1/27. For step 2 and onward, the volume U of each cube removed is 1/27 the volume of each cube removed in the previous step. The volume V of the sponge at any particular step is simply the difference of the previous value of V (the volume at the previous step) and the product of C and U (the volume removed during the current step).

(b) Since $20/27 < 1$, the Menger Sponge has infinitesimally small volume. (If a number between 0 and 1 is multiplied by itself over and over again, the resulting product will be close to 0.)

51. If it is attracted to a number, then we have $s_{N+1} = (s_N)^2 + (-0.75)$ and $s_{N+1} = s_N$. Substituting, we have

$s_N = (s_N)^2 - 0.75$ so $s_N^2 - s_N - 0.75 = 0$. Solve via the quadratic equation: $s_N = \dfrac{1 \pm \sqrt{1 - 4\left(-\frac{3}{4}\right)}}{2}$.

Solving yields $s_N = -\dfrac{1}{2}$ or $s_N = \dfrac{3}{2}$. Examining the first few terms of the sequence indicates that the

sequence is attracted to $-\dfrac{1}{2}$.

53. The first twenty steps: 0.3125, -1.15234375, 0.0778961182, -1.24393219, 0.297367305, -1.16157269, 0.0992511044, -1.24014922, 0.287970084, -1.16707323, 0.112059926, -1.23744257, 0.281264121, -1.17089049, 0.120984549, -1.23536274, 0.276121097, -1.17375714, 0.127705824, -1.23369122. The numbers oscillate, alternating between positive and negative values, but always staying within the interval $-1.25 \le x \le 0.3125$. The sequence is periodic.

RUNNING

55. The critical observation is the following: Suppose that P is an arbitrary point inside (or on the boundary) of triangle AM_1M_2. Then the midpoint Q of the segment PB is inside (or on the boundary) of triangle BM_1M_3. (Why? Because segment M_1M_3 is parallel to AC and therefore Y is the midpoint of segment BX, which forces Q to be between Y and B.) The argument above implies that the chaos game will force us to keep hopping among the insides (or boundaries) of the triangles AM_1M_2, BM_1M_3, and CM_2M_3, but never into a point inside triangle $M_1M_2M_3$.

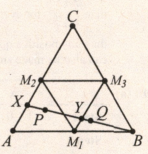

57. (a) The first twenty steps: $-0.25 + 0.125i$, $-0.203125 + 0.1875i$, $-0.243896484 + 0.173828125i$, $-0.220730722 + 0.165207863i$, $-0.228571586 + 0.177067098i$, $-0.229107787 + 0.169054985i$, $-0.22608921 + 0.172536373i$, $-0.228652469 + 0.171982776i$, $-0.227296123 + 0.171351427i$, $-0.227697784 + 0.17210497i$, $-0.22777384 + 0.17162416i$, $-0.22757393 + 0.171817012i$, $-0.227731192 + 0.171797854i$, $-0.227653007 + 0.17175254i$, $-0.227673043 + 0.171800036i$, $-0.227680238 + 0.171771526i$, $-0.227667167 + 0.171782036i$, $-0.227676729 + 0.171781741i$, $-0.227672274 + 0.17177859i$, $-0.22767322 + 0.171781556i$.
The sequence is attracted to $-0.22767 + 0.17178i$ (rounded to 5 decimal places).

(b) The first twenty steps: $-0.25 - 0.125i$, $-0.203125 - 0.1875i$, $-0.243896484 - 0.173828125i$, $-0.220730722 - 0.165207863i$, $-0.228571586 - 0.177067098i$, $-0.229107787 - 0.169054985i$, $-0.22608921 - 0.172536373i$, $-0.228652469 - 0.171982776i$, $-0.227296123 - 0.171351427i$, $-0.227697784 - 0.17210497i$, $-0.22777384 - 0.17162416i$, $-0.22757393 - 0.171817012i$, $-0.227731192 - 0.171797854i$, $-0.227653007 - 0.17175254i$, $-0.227673043 - 0.171800036i$, $-0.227680238 - 0.171771526i$, $-0.227667167 - 0.171782036i$, $-0.227676729 - 0.171781741i$, $-0.227672274 - 0.17177859i$, $-0.22767322 - 0.171781556i$.
The sequence is attracted to $-0.22767 - 0.17178i$ (rounded to 5 decimal places).

59. The Koch curve consists of $N = 4$ self-similar copies of itself with each copy having been reduced by a scaling factor of $S = 3$. So, the dimension of the Koch curve is $D = \dfrac{\log 4}{\log 3} \approx 1.26$.

Chapter 13

13.1 Fibonacci Numbers

1. (a) $F_{15} = 610$ (the 15th Fibonacci number)

   (b) $F_{15} - 2 = 610 - 2 = 608$

   (c) $F_{15-2} = F_{13} = 233$

   (d) $\dfrac{F_{15}}{5} = \dfrac{610}{5} = 122$

   (e) $F_{15/5} = F_3 = 2$

3. (a) $F_1 + F_2 + F_3 + F_4 + F_5 = 1+1+2+3+5$
   $= 12$

   (b) $F_{1+2+3+4+5} = F_{15} = 610$

   (c) $F_3 \times F_4 = 2 \times 3 = 6$

   (d) $F_{3\times4} = F_{12} = 144$

5. (a) $3F_N + 1$ represents one more than three times the Nth Fibonacci number.

   (b) $3F_{N+1}$ represents three times the Fibonacci number in position $(N+1)$.

   (c) $F_{3N} + 1$ represents one more than the Fibonacci number in the $3N$th position.

   (d) F_{3N+1} represents the Fibonacci number in position $(3N+1)$.

7. (a) $F_{38} = F_{37} + F_{36}$
   $= 24,157,817 + 14,930,352$
   $= 39,088,169$

   (b) $F_{39} = F_{38} + F_{37}$
   $= 39,088,169 + 24,157,817$
   $= 63,245,986$

9. (a) $F_{35} = F_{37} - F_{36}$
   $= 24,157,817 - 14,930,352$
   $= 9,227,465$

   (b) $F_{34} = F_{36} - F_{35}$
   $= 14,930,352 - 9,227,465$
   $= 5,702,887$

11. Binet's simplified formula is

$$F_N = \left[\!\left[\left(\frac{1+\sqrt5}{2}\right)^N \middle/ \sqrt5 \right]\!\right] \text{ where } [\![\;]\!] \text{ denotes}$$

'round to the nearest integer.' So,

$$F_{20} = \left[\!\left[\left(\frac{1+\sqrt5}{2}\right)^{20} \middle/ \sqrt5 \right]\!\right]. \text{ To evaluate this}$$

expression, first compute $\dfrac{1+\sqrt5}{2}$, then raise that value to the 20th power, then divide the result by $\sqrt5$. This gives 6764.999964. Since $[\![6764.999964]\!] = 6765$, we have that $F_{20} = 6765$.

13. (a) Fifth equation in the sequence:
   $1 + 2 + 5 + 13 + 34 + 89 = 144$
   (In words, the left hand side is the sum of every other Fibonacci number. The right hand side is the next Fibonacci after the last one appearing on the left hand side.)

   (b) 22. In each case the Fibonacci number that appears on the right of the equality is that Fibonacci number immediately following the largest Fibonacci number appearing on the left.

   (c) $N+1$. $N+1$ is one bigger than N.

15. (a) $F_8 \cdot F_{11} = 21 \cdot 89 = 1869$. But we also have $F_{10}^2 - F_9^2 = 55^2 - 34^2 = 1869$.

   (b) $F_N \cdot F_{N+3} = F_{N+2}^2 - F_{N+1}^2$

17. (a) Here we are adding the $(N+1)$st and the $(N+2)$nd Fibonacci numbers. The result is the $(N+3)$rd Fibonacci number. That is,
   $F_{N+1} + F_{N+2} = F_{N+3}$.

   (b) Since $F_{N-2} + F_{N-1} = F_N$, we have
   $F_N - F_{N-2} = F_{N-1}$.

(c) Add two terms at a time.

$$F_N + F_{N+1} + F_{N+3} + F_{N+5}$$
$$= (F_N + F_{N+1}) + F_{N+3} + F_{N+5}$$
$$= (F_{N+2}) + F_{N+3} + F_{N+5}$$
$$= (F_{N+2} + F_{N+3}) + F_{N+5}$$
$$= F_{N+4} + F_{N+5}$$
$$= F_{N+6}$$

19. (a)
$$1 + \frac{F_N}{F_{N-1}} = \frac{F_{N-1}}{F_{N-1}} + \frac{F_N}{F_{N-1}}$$
$$= \frac{F_{N-1} + F_N}{F_{N-1}}$$
$$= \frac{F_{N+1}}{F_{N-1}}$$

(b)
$$\frac{F_{N-1}}{F_N} - 1 = \frac{F_{N-1}}{F_N} - \frac{F_N}{F_N}$$
$$= \frac{F_{N-1} - F_N}{F_N}$$
$$= \frac{-(F_N - F_{N-1})}{F_N}$$
$$= -\frac{F_{N-2}}{F_N}$$

21. (a) $A_1 = 5$, $A_2 = 5$,
$$A_3 = A_1 + A_2 = 5 + 5 = 10,$$
$$A_4 = A_2 + A_3 = 5 + 10 = 15,$$
$$A_5 = A_3 + A_4 = 10 + 15 = 25,$$
$$A_6 = A_4 + A_5 = 15 + 25 = 40,$$
$$A_7 = A_5 + A_6 = 25 + 40 = 65,$$
$$A_8 = A_6 + A_7 = 40 + 65 = 105,$$
$$A_9 = A_7 + A_8 = 65 + 105 = 170,$$
$$A_{10} = A_8 + A_9 = 105 + 170 = 275$$

(b) Note that every term in the sequence found in (a) is 5 times as large as the corresponding term in the Fibonacci sequence (i.e. $A_1 = 5 \cdot F_1$, $A_2 = 5 \cdot F_2$, $A_3 = 5 \cdot F_3$, etc.). So, $A_{25} = 5 \cdot F_{25}$ $= 5 \times 75,025 = 375,125$.

(c) $A_N = 5 \cdot F_N$

23. (a) $L_1 = 1$, $L_2 = 3$,
$$L_3 = L_1 + L_2 = 1 + 3 = 4,$$

$$L_4 = L_2 + L_3 = 3 + 4 = 7,$$
$$L_5 = L_3 + L_4 = 4 + 7 = 11,$$
$$L_6 = L_4 + L_5 = 7 + 11 = 18,$$
$$L_7 = L_5 + L_6 = 11 + 18 = 29,$$
$$L_8 = L_6 + L_7 = 18 + 29 = 47,$$
$$L_9 = L_7 + L_8 = 29 + 47 = 76,$$
$$L_{10} = L_8 + L_9 = 47 + 76 = 123,$$
$$L_{11} = L_9 + L_{10} = 76 + 123 = 199,$$
$$L_{12} = L_{10} + L_{11} = 123 + 199 = 322 .$$

(b) $N = 1: L_1 = 1 = 2 \times 1 - 1 = 2F_2 - F_1$;
$N = 2: L_2 = 3 = 2 \times 2 - 1 = 2F_3 - F_2$;
$N = 3: L_3 = 4 = 2 \times 3 - 2 = 2F_4 - F_3$;
$N = 4: L_4 = 7 = 2 \times 5 - 3 = 2F_5 - F_4$.

(c) $L_{20} = 2F_{21} - F_{20}$
$$= 2 \times 10,946 - 6765$$
$$= 15,127$$

13.2 The Golden Ratio

25. (a) First, we put $x^2 = x + 1$ into the standard form $x^2 - x - 1 = 0$. The solutions to this quadratic equation are then

$$x = \frac{-(-1) \pm \sqrt{(-1)^2 - 4(1)(-1)}}{2(1)}$$
$$= \frac{1 \pm \sqrt{5}}{2}$$

So,

$$x = \frac{1 + \sqrt{5}}{2} \quad \text{or} \quad x = \frac{1 - \sqrt{5}}{2}$$
$$\approx 1.61803 \qquad\qquad \approx -0.61803$$

(b) $\dfrac{1 + \sqrt{5}}{2} + \dfrac{1 - \sqrt{5}}{2} = 1$

(c) One solution is positive, the second solution is negative, and their sum equals 1. This implies that when you add the two solutions their decimal parts must cancel each other out.

27. (a) First, we put $3x^2 = 8x + 5$ into the standard form $3x^2 - 8x - 5 = 0$. The solutions to this quadratic equation are then

$$x = \frac{-(-8) \pm \sqrt{(-8)^2 - 4(3)(-5)}}{2(3)}$$

$$= \frac{8 \pm \sqrt{124}}{6}$$

$$= \frac{4 \pm \sqrt{31}}{3}$$

So, $x = \frac{4 + \sqrt{31}}{3} \approx 3.18925$, or

$x = \frac{4 - \sqrt{31}}{3} \approx -0.52259$

(b) $\frac{4+\sqrt{31}}{3} + \frac{4-\sqrt{31}}{3} = \frac{8}{3}$

29. (a) Substituting $x = 1$ into the equation gives $55(1)^2 = 34(1) + 21$. Since the equation is satisfied, $x = 1$ is a solution.

(b) Rewrite the equation:
$55x^2 - 34x - 21 = 0$
Then $a = 55$, $b = -34$, and $1 + x = \frac{34}{55}$ so

that $x = \frac{34}{55} - 1 = -\frac{21}{55} \approx -0.38182$.

31. (a) Substituting $x = -1$ into the equation gives $21 = 34(-1) + 55$ (i.e. that 21 is the difference of the two subsequent Fibonacci numbers). Since the equation is satisfied, $x = -1$ is a solution.

(b) Rewrite the equation:
$21x^2 - 34x - 55 = 0$
Then $a = 21$, $b = -34$, and $-1 + x = \frac{34}{21}$ so

that $x = \frac{34}{21} + 1 = \frac{55}{21} \approx 2.61905$.

33. (a) Substituting $x = 1$ in the equation gives $F_N = F_{N-1} + F_{N-2}$ which is a defining equation for the Fibonacci numbers. Therefore, the equation is satisfied and $x=1$ is, by definition, a solution.

(b) Rewrite the equation:
$F_N x^2 - F_{N-1}x - F_{N-2} = 0$

Then, using the hint in Exercise 31(b),

$1 + x = \frac{F_{N-1}}{F_N}$ so that

$$x = \left(\frac{F_{N-1}}{F_N}\right) - 1$$

$$= \frac{F_{N-1} - F_N}{F_N}$$

$$= \frac{-(F_N - F_{N-1})}{F_N}$$

$$= \frac{-F_{N-2}}{F_N}.$$

35. (a) $\frac{1}{\phi} \approx 0.6180339887$

(b) Start with the golden property: $\phi^2 = \phi + 1$.

Divide both sides by ϕ and get $\phi = 1 + \frac{1}{\phi}$.

It follows that ϕ and ϕ must have exactly the same decimal part (they are 1 unit apart).

37. (a) $F_{500} \approx \phi \times F_{499}$

$\approx 1.6180 \times 8.617 \times 10^{103}$

$\approx 13.94 \times 10^{103}$

$= 1.394 \times 10^{104}$

Remember that when using scientific notation, only one digit may occur before the decimal point.

(b) $F_{498} \approx \frac{F_{499}}{\phi}$

$\approx \frac{8.617 \times 10^{103}}{1.618}$

$\approx 5.326 \times 10^{103}$

39. (a) $A_7 = 2A_{7-1} + A_{7-2}$

$= 2A_6 + A_5$

$= 2(70) + 29$

$= 169$

(b) $\frac{A_7}{A_6} = \frac{169}{70}$

≈ 2.41429

(c) $\dfrac{A_{11}}{A_{10}} = \dfrac{5,741}{2,378}$

≈ 2.41421

(d) 2.41421 (Though not obvious, the ratio actually approaches $1+\sqrt{2}$ as the value of *N* increases.)

13.3 Gnomons

41. (a) Since *R* and *R'* are similar, each side length of *R'* is 3 times as long as the corresponding side in *R*. So, the perimeter of *R'* will be 3 times as great as the perimeter of *R*. This means that the perimeter of *R'* is $3 \times 41.5 = 124.5$ inches.

(b) Since each side of *R'* is 3 times as long as the corresponding side in *R*, the area of *R'* will be $3^2 = 9$ times as larges as the area of *R*. That is, the area of *R'* is $9 \times 105 = 945$ square inches.

43. (a) The ratio of the perimeters must be the same as the ratio of corresponding side lengths.

$\dfrac{P}{13 \text{ in.}} = \dfrac{60 \text{ m}}{5 \text{ in.}}$

$P = \dfrac{60 \text{ m}}{5 \text{ in.}} \times 13 \text{ in.}$

$= 156 \text{ m}$

(b) We use the fact that the ratio of the areas of two similar triangles (or any two similar polygons) is the same as the ratio of their side lengths squared. So,

$\dfrac{A}{20 \text{ in.}^2} = \left(\dfrac{60 \text{ m}}{5 \text{ in.}}\right)^2$

$A = \left(\dfrac{60 \text{ m}}{5 \text{ in.}}\right)^2 \times (20 \text{ in.}^2)$

$= 2880 \text{ m}^2$

45. The two rectangles that need to be similar in order for the shaded rectangle to be a gnomon are the 2 by 10 rectangle and the 10 by *x*+2 rectangle (the one found by 'gluing' the white rectangle and the shaded rectangle together). The ratios of the sides of these rectangles must be the same.

$\dfrac{10}{2} = \dfrac{x+2}{10}$

$2(x+2) = 100$

$2x+4 = 100$

$2x = 96$

$x = 48$

47. The two rectangles that need to be similar in order for the shaded ring to be a gnomon are the 8 by 12 white rectangle and the 12 by 14+*x* rectangle (the one found by 'gluing' the white rectangle and the shaded ring together). The ratios of the sides of these rectangles must be the same.

$\dfrac{8}{12} = \dfrac{3+8+1}{2+12+x}$

$\dfrac{8}{12} = \dfrac{12}{14+x}$

$8(14+x) = 144$

$112+8x = 144$

$8x = 32$

$x = 4$

49. The two rectangles that need to be similar in order for the shaded shape to be a gnomon are the 5 by *x* white rectangle and the 9 by 4+*x* rectangle (the one found by 'gluing' the white rectangle and the shaded ring together). The ratios of the sides of these rectangles must be the same.

$\dfrac{5}{x} = \dfrac{5+2+2}{x+2+2}$

$\dfrac{5}{x} = \dfrac{9}{x+2}$

$5x+20 = 9x$

$20 = 4x$

$x = 5$

51. (a) In the figure, the measure of angle CAD must be $180° - 108° = 72°$. Since triangle BDC must be similar to triangle BCA (in order for triangle ACD to be a gnomon), it must be that the measure of angle BDC must be $36°$. Using the fact that the sum of the measures of the angles in any triangle is $180°$, it follows that the measure of angle ACD is $72°$.

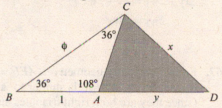

(b) Since triangle DBC is isosceles, $x = \phi$. Note that triangle ACD is also isosceles. Hence, $y = x = \phi$ as well. Likewise, triangle ABC is isosceles so that the length of AC is 1.

53. The two triangles that need to be similar in order for the shaded shape to be gnomon are the 3-4-5 white triangle and the 9-x-(5+y) triangle (the one found by 'gluing' the white triangle and the shaded region together). By comparing side lengths of the similar triangles, we find $\dfrac{3}{9} = \dfrac{4}{x}$ and $\dfrac{3}{9} = \dfrac{5}{5+y}$. That is, $3x=36$ and $3(5+y) = 45$. Solving these gives $x = 12$ and $y = 10$.

JOGGING

55. (a) $\dfrac{F_1}{F_2} = \dfrac{1}{1} = 1$, $\dfrac{F_2}{F_3} = \dfrac{1}{2} = 0.5$,

$\dfrac{F_3}{F_4} = \dfrac{2}{3} = 0.666667$, $\dfrac{F_4}{F_5} = \dfrac{3}{5} = 0.6$,

$\dfrac{F_5}{F_6} = \dfrac{5}{8} = 0.625$, $\dfrac{F_6}{F_7} = \dfrac{8}{13} = 0.615385$,

$\dfrac{F_7}{F_8} = \dfrac{13}{21} = 0.619048$, 0.617647,

0.618182, 0.61798, 0.618056, 0.618026, 0.618037, 0.618033, ...

(b) We know that $\dfrac{F_{N+1}}{F_N} \to \phi$ as N gets large. It follows that the reciprocals of these ratios approaches the reciprocal of this value (i.e. that $\dfrac{F_N}{F_{N+1}} \to \dfrac{1}{\phi}$). Since $\dfrac{1}{\phi} = \phi - 1$, we have the result. [The fact that $\dfrac{1}{\phi} = \phi - 1$ follows from using the golden rule $\phi^2 = \phi + 1$ and dividing both sides by ϕ.]

57. (a) $T_1 = 1$, $T_2 = 1 + \dfrac{1}{T_1} = 1 + \dfrac{1}{1} = 2$,

$T_3 = 1 + \dfrac{1}{T_2} = 1 + \dfrac{1}{2} = \dfrac{3}{2}$,

$T_4 = 1 + \dfrac{1}{T_3} = 1 + \dfrac{1}{(3/2)} = 1 + \dfrac{2}{3} = \dfrac{5}{3}$,

$T_5 = 1 + \dfrac{1}{T_4} = 1 + \dfrac{1}{(5/3)} = 1 + \dfrac{3}{5} = \dfrac{8}{5}$,

$T_6 = 1 + \dfrac{1}{T_5} = 1 + \dfrac{1}{(8/5)} = 1 + \dfrac{5}{8} = \dfrac{13}{8}$.

(b) We know that $\dfrac{F_{N+1}}{F_N} \to \phi$ as N gets large. But $T_N = \dfrac{F_{N+1}}{F_N}$ from part (a).

59. (a) $K_1 = 2F_2 - F_1 = 2 \times 1 - 1 = 1$,

$K_2 = 2F_3 - F_2 = 2 \times 2 - 1 = 3$, and

$K_{N-1} + K_{N-2} = (2F_N - F_{N-1}) + (2F_{N-1} - F_{N-2})$
$= 2(F_N + F_{N-1}) - (F_{N-1} + F_{N-2})$
$= 2F_{N+1} - F_N$
$= K_N$

Since the seeds are the same and the recursive rule is the same, it follows that $K_N = L_N$.

(b) From part (a), we have that

$$\frac{L_{N+1}}{L_N} = \frac{2F_{N+2} - F_{N+1}}{2F_{N+1} - F_N}.$$ If you divide the

numerator and the denominator of the right hand expression by F_{N+1} you get

$$\frac{L_{N+1}}{L_N} = \frac{2F_{N+2} - F_{N+1}}{2F_{N+1} - F_N}$$

$$= \frac{(2F_{N+2} - F_{N+1})/F_{N+1}}{(2F_{N+1} - F_N)/F_{N+1}}$$

$$= \frac{2(F_{N+2}/F_{N+1}) - 1}{2 - (F_N/F_{N+1})}$$

$$\rightarrow \frac{2\phi - 1}{2 - (1/\phi)}$$

$$= \frac{\phi(2\phi - 1)}{2\phi - 1}$$

$$= \phi$$

since $\dfrac{F_{N+1}}{F_N} \rightarrow \phi$.

61. First, remember that the sum of two odd numbers is always even and the sum of an odd number and an even number is always odd. Now, since the seeds in the Fibonacci sequence (1 and 1) are both odd, the third number is even. Since the second number is odd and the third number is even, the fourth number is odd. Similarly, since the third number is even and the fourth number is odd, the fifth number is odd. But then the sixth number, the sum of two odd numbers again like the seeds, is even. It follows that every third number is even and the others in the sequence are all odd.

A nice way to visualize this and convince yourself of its truth is the following:
O O E O O E O O E O O E O O E O O E …

63.

```
      A      B
20  ┌────┬──────┐
    │    │      │
    │ 10 │  x   │
    └────┴──────┘
      10 + x
```

$$\frac{10}{20} = \frac{20}{10 + x}$$

$$10(10 + x) = 400$$

$$100 + 10x = 400$$

$$10x = 300$$

So, $x = 30$. Rectangle B is 20 by 30.

65. Yes, but only when $x = y$ (i.e. if the picture is a square). The reason is that for a frame to be a gnomon to the picture, we must have $(x+2w)/(y+2w) = x/y$. Solving for x in terms of y gives $x = y$.

$$\frac{x + 2w}{y + 2w} = \frac{x}{y}$$

$$y(x + 2w) = x(y + 2w)$$

$$xy + 2wy = xy + 2wx$$

$$2wy = 2wx$$

$$y = x$$

67. From elementary geometry $\angle AEF = \angle DBA$. Consequently $\blacktriangle AEF$ is similar to $\blacktriangle DBA$ (since they are both right triangles and so have all their corresponding angles congruent). So $AF : FE = DA : AB$ which shows rectangle $ADEF$ is similar to rectangle $ABCD$.

69. Follows from exercise 66(a) and the following figure.

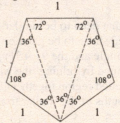

RUNNING

71. (a) Using the same seeds as G_N (namely a and b), we define a sequence $K_N = bF_{N-1} + aF_{N-2}$ and show that the numbers K_N satisfy the same rule as G_N (namely that $G_N = G_{N-1} + G_{N-2}$). But,

$$K_{N-1} + K_{N-2} = (bF_{N-2} + aF_{N-3}) + (bF_{N-3} + aF_{N-4})$$

$$= b(F_{N-2} + F_{N-3}) + a(F_{N-3} + F_{N-4})$$

$$= bF_{N-1} + aF_{N-2}$$

$$= K_N$$

It follows that $K_N = G_N$.

(b) From part (a), we have that

$\dfrac{G_{N+1}}{G_N} = \dfrac{bF_N + aF_{N-1}}{bF_{N-1} + aF_{N-2}}$. If you divide the

numerator and the denominator of the right hand expression by F_{N-1} you get

$$\dfrac{G_{N+1}}{G_N} = \dfrac{bF_N + aF_{N-1}}{bF_{N-1} + aF_{N-2}}$$

$$= \dfrac{\left(bF_N + aF_{N-1}\right)/F_{N-1}}{\left(bF_{N-1} + aF_{N-2}\right)/F_{N-1}}$$

$$= \dfrac{b\left(F_N/F_{N-1}\right) + a}{b + a\left(F_{N-2}/F_{N-1}\right)}$$

$$\rightarrow \dfrac{b\phi + a}{b + a(1/\phi)}$$

$$= \dfrac{\phi(b\phi + a)}{b\phi + a}$$

$$= \phi$$

since $\dfrac{F_{N+1}}{F_N} \rightarrow \phi$.

73. $F_1 = F_3 - F_2$

$F_2 = F_4 - F_3$

$\vdots$

$F_{N-1} = F_{N+1} - F_N$

$F_N = F_{N+2} - F_{N+1}$

Adding the left and right hand sides respectively, yields

$$F_1 + F_2 + \cdots + F_N = F_{N+2} - F_2 = F_{N+2} - 1$$

75. Suppose M is a positive integer greater than 2. If M is a Fibonacci number ($M = F_N$), then it is certainly true that M can be written as the sum of distinct Fibonacci numbers ($M = F_N = F_{N-1} + F_{N-2}$). Otherwise, M falls somewhere between two Fibonacci numbers, say between F_N and F_{N+1}. Write $M = F_N + (M - F_N)$. Now $M - F_N < F_{N+1} - F_N = F_{N-1}$ is itself a positive integer smaller than M and we can repeat the argument. Eventually the process must stop. (This is an argument by induction in disguise.)

Note: This argument gives an algorithm for finding a decomposition of M as a sum of Fibonacci numbers. Find the largest Fibonacci number that fits into M (call it F_N), then find the largest Fibonacci number that fits into the difference $M - F_N$, and so on.

77. (a) From exercise 68, $x = \phi y$ and so $\dfrac{x}{y} = \phi$.

(See figure.) Since $z = x$,

$$\dfrac{x+y}{z} = \dfrac{x+y}{x} = 1 + \dfrac{y}{x} = 1 + \dfrac{1}{\phi} = \phi.$$

Therefore, $\dfrac{x+y+z}{x+y} = 1 + \dfrac{z}{x+y} =$

$1 + \dfrac{1}{\phi} = \phi.$

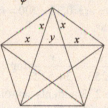

(b) From part (a), $x = \phi y = \phi$;

$x + y = z\phi = x\phi = \phi^2$;

$x + y + z = (x+y)\phi$

$\qquad\qquad = \phi^2 \cdot \phi$

$\qquad\qquad = \phi^3$

Chapter 14

14.1 Enumeration

1. 64; the proportion of red gumballs in the sample is $\frac{8}{25} = 0.32$. Assuming that the sample is representative of the population, we estimate that there are $0.32 \times 200 = 64$ red gumballs in the jar.

3. 2320; we solve $\frac{17}{253} = \frac{x}{34,522}$ for x. The proportion of the 253 patients sampled having A- blood type should be the same as the proportion of the 34,522 people in Madison County having A- blood type.

5. 2549; we solve $\frac{121}{200} = \frac{1542}{x}$ for x. The proportion of the 200 seats sampled that were female should be the same as the proportion of the x people at the concert that were female.

7. 2000; we solve $\frac{30}{120} = \frac{500}{N}$ for N. The proportion of the 120 fish recaptured that had tags ($k = 30$) should be the same as the proportion of the 500 fish in the pond that were tagged during the first capture.

9. 274; we solve $\frac{76}{172} = \frac{121}{N}$ for N. The proportion of the 172 whales recaptured in 2008 that had tags ($k = 76$) should be the same as the proportion of the 121 whales that were tagged during the first (photographic) capture in 2007.

11. We now solve $\frac{77}{173} = \frac{122}{N+1}$ for N. This gives $N = \frac{122 \times 173}{77} - 1 = 273$ which is one less than under the standard capture-recapture formula.

13. Using $\frac{k}{n_2} = \frac{n_1}{N}$ with $k = 7, n_1 = 1700$, and $N = 160,000$, we solve $\frac{7}{n_2} = \frac{1700}{160,000}$ for n_2 to find $n_2 \approx 660$.

14.2 Measurement

15. (a) B - Convenience sampling.
George is selecting those units in the population that are easily accessible.

(b) D - Stratified sampling.
The student newspaper is dividing the population into strata and then selecting a proportionately sized random sample from each stratum (in fact, 5% from each stratum).

(c) A - Simple random sampling.
Every subset of three players has the same chance of being selected as any other subset of three players.

(d) C - Quota sampling.
The coach is attempting to force the sample to fit a particular profile.

17. (a) The registered voters in Cleansburg.

(b) The 680 registered voters that are polled by telephone.

(c) Simple random sampling.

19. Of the people surveyed, 45% indicated they would vote for Smith $\left(\frac{306}{680} = 0.45 \right)$. The actual percentage was 42%. Since 45% − 42% = 3%, the sampling error for Smith was 3%. Similarly, the sampling error for Jones was 43% - 40% = 3% and the sampling error for Brown was 15% - 15% = 0%.

21. (a) The sample for this survey is the 350 students attending the Eureka High football game before the election.

(b) $\frac{350}{1250} = 28\%$

23. (a) The population consists of all 1250 students at Eureka High School while the sampling frame consists only of those students that attended the football game a week prior to the election.

(b) The sampling error is mainly a result of sampling bias. The sampling frame (and hence any sample taken from it) is not representative of the population. Students that choose to attend a football game are not representative of all Eureka High students.

25. (a) The target population of this survey is the citizens of Cleansburg.

(b) The sampling frame is limited to that part of the target population that passes by a city street corner between 4:00 p.m. and 6:00 p.m.. It excludes citizens of Cleansburg having other responsibilities during that time of day.

27. (a) The choice of street corner could make a great deal of difference in the responses collected.

(b) *D* (We are making the assumption that people who live or work downtown are much more likely to answer yes than people in other parts of town.)

(c) Yes, the survey was subject to selection bias for two main reasons. (i) People out on the street between 4 p.m. and 6 p.m. are not representative of the population at large. For example, office and white-collar workers are much more likely to be in the sample than homemakers and school teachers. (ii) The five street corners were chosen by the interviewers and the passersby are unlikely to represent a cross section of the city.

(d) No, no attempt was made to use quotas to get a representative cross section of the population.

29. (a) Stratified sampling. The trees are broken into three different strata (by variety) and then a random sample is taken from each stratum.

(b) Quota sampling. The grower is using a systematic method to force the sample to fit a particular profile. However, because the grower is human, sampling bias could then be introduced. When selecting 300 trees of variety A, the grower does not select them at random. Selecting 300 trees in one particular part of the orchard could bias the yield.

31. (a) Label the crates with numbers 1 to 250. Select 20 of these numbers at random (put the 250 numbers in a hat and draw out 20). Sample those 20 crates. The important point is that *any* set of 20 crates have an equal chance of being in the sample.

(b) Select the top 20 crates in the shipment (those easiest to access).

(c) Randomly sample 6 crates from supplier A, 6 crates from supplier B, and 8 crates from supplier C.

(d) Select any 6 crates from supplier A, any 6 crates from supplier B, and any 8 crates from supplier C.

14.3 Cause and Effect

33. (a) The target population for this study is anyone who could have a cold and would consider buying vitamin *X* (i.e., pretty much all adults).

(b) The sampling frame is only a small portion of the target population. It only consists of college students in the San Diego area that are suffering from colds.

(c) The sample is 500 of these San Diego area students all of whom are suffering from a cold at the start of the study. Selection bias is present here since this sample would likely under represent older adults and those living in colder climates.

35. Four different problems with this study that indicate poor design include
(i) using college students (College students are not a representative cross section of the population in terms of age and therefore in terms of how they would respond to the treatment.),
(ii) using subjects only from the San Diego area,
(iii) offering money as an incentive to participate, and
(iv) allowing self-reporting (the subjects themselves determine when their colds are over).

37. (a) The target population for this study is all potential knee surgery patients.

    (b) The sample consists of 180 potential knee surgery patients at the Houston VA Medical Center who volunteered to be in the study.

39 (a) Yes, this study could be considered a randomized controlled experiment since the 180 patients in the study were assigned to a treatment group at random.

    (b) This was a blind experiment. The doctors certainly knew which surgery they were performing on each patient.

41. The target population consists of all people having a history of colorectal adenomas. The point of the study is to determine the effect that Vioxx had on this population. That is, whether recurrence of colorectal polyps could be prevented in this population.

43. (a) The treatment group consisted of the 1287 patients that were given 25 daily milligrams of Vioxx. The control group consisted of the 1299 patients that were given a placebo.

    (b) This is an experiment since members of the population received a treatment. It is a controlled placebo experiment since there was a control group that did not receive the treatment, but instead received a placebo. It is a randomized controlled experiment since the 2586 participants were randomly divided into the treatment and control groups. The study was double blind since neither the participants nor the doctors involved in the clinical trial knew who was in each group.

45. (a) Women; particularly young women

    (b) The sampling frame consists of those women between 16 and 23 years of age that are not at high risk for HPV infection (that is, those women having no prior abnormal pap smears and at most five previous male sexual partners). Pregnant women are also excluded from the sampling frame due to the risks involved. Though not perfect, it appears to be a representative subset of the target population.

47. (a) The treatment group consists of the half of the 2392 women in the sample that received the HPV vaccine.

    (b) This is a controlled placebo experiment since there was a control group that did not receive the treatment, but instead received a placebo injection. It is a randomized controlled experiment since the 2392 participants were randomly divided into the treatment and control groups. The study was likely double blind – it is probable that neither the participants nor the medical personnel giving the injections knew who was in each group.

49. (a) The stated purpose of the study was to determine the effectiveness of a new serum for treating diphtheria. So, the target population consisted of all people suffering from diphtheria.

    (b) The sampling frame consists of those people having diphtheria symptoms serious enough to be admitted to one particular Copenhagen hospital (between May 1896 and May 1897). It appears to be a reasonably representative subset of the target population. This is primarily because there is likely little difference between those suffering from diphtheria in this Copenhagen hospital and those suffering from diphtheria elsewhere.

51. (a) The treatment group consisted of those patients admitted on the "even" days that received both the new serum and the standard treatment. The control group consisted of those patients admitted on the "odd" days that received only the standard treatment for diphtheria at the time.

    (b) When received together with standard treatment, this new serum appears to be somewhat effective at treating diphtheria. Significantly more patients that did not receive the treatment in this study died (presumably many of them from diphtheria or related issues).

53. (a) The target population consisted of men with prostate enlargement; particularly older men.

(b) The sampling frame consists of men aged 45 or older having moderately impaired urinary flow. It appears to be a representative subset of the target population.

55. (a) The treatment group consisted of the group of volunteers that received saw palmetto daily.

(b) The saw palmetto odor would be a dead giveaway as to who was getting the treatment and who was getting a placebo. To make the study truly blind, the odor had to be covered up.

(c) Yes. There was a control group that received a placebo instead of the treatment, and the treatment group was randomized. We can infer the study was blind from the fact that an effort was made to hide the odor of the saw palmetto. From the description of the study there is no way to tell if the study was also double blind.

JOGGING

57. (a) Assuming that the registrar has a complete list of the 15,000 undergraduates at Tasmania State University, the target population and the sampling frame both consist of all undergraduates at TSU.

(b) $N = 15,000$

59. (a) In simple random sampling, any two members of the population have as much chance of both being in the sample as any other two. But in this sample, two people with the same last name—say Len Euler and Linda Euler—have no chance of being in the sample together.

(b) Sampling variability. The students sampled appear to be a reasonably representative cross section of all TSU undergraduates that might or might not be familiar with the new financial aid program.

61. To estimate the number of quarters, we disregard the nickels and dimes—they are irrelevant. To that end, we solve $\frac{4}{28} = \frac{36}{N}$ for N to estimate the number of quarters in the jar. This gives $N = 252$ quarters.

Similarly, to estimate the number of nickels, we disregard the quarters and dimes. In particular, we solve $\frac{5}{29} = \frac{45}{N}$ for N to estimate the number of nickels in the jar. This gives $N = 261$ nickels.

Lastly, we solve $\frac{8}{43} = \frac{69}{N}$ for N to estimate the number of dimes in the jar. This gives approximately $N = 371$ nickels.

Since 252 quarters, 261 nickels, and 371 dimes are estimated to be in the jar, the total amount of money in the jar is estimated to be $252 \times (\$0.25) + 261 \times (\$0.05) + 371 \times (\$0.10) = \$63.00 + \$13.05 + \$37.10 = \$113.15$

63. Capture-recapture would overestimate the true population. If the fraction of those tagged in the recapture appears lower than it is in reality, then the fraction of those tagged in the initial capture will also be computed as lower than the truth. This makes the total population appear larger than it really is.

For an example to see how this works, suppose a population had $N = 1000$ individuals. Imagine an initial tagging of $n_1 = 100$ individuals (10% of the population are tagged). Then, in the recapture of $n_2 = 20$, we should expect $k = 2$ individuals are tagged. If, however, some of the individuals were more likely to be prey, then the number of individuals that are captured twice (the value of k) would be lower than expected (less than 2). Say $k = 1$ individuals (half the original number) were captured twice. Then, the value of N would be estimated to be

$N = \frac{n_2}{k} \times n_1 = \frac{20}{1} \times 100 = 2000$. This is clearly an overestimate (twice the original N-value).

65. Answers will vary. For example, Al might see that Apple's stock price was down today and infer that the entire stock market had a down day. Or Betty might look at the headlines in the New York Times and determine that her son stationed overseas must be safe and sound at his post. Or Carla might go to the mall on Black Friday and deem it always that busy.

67. Consideration of one exception made them less likely to consider a second exception. This is why it is critical that the order a series of survey questions such as this are asked is randomized.

69. (a) An Area Code 900 telephone poll represents an extreme case of selection bias. People that respond to these polls usually represent the extreme view points (strongly for or strongly against), leaving out much of the middle of the road point of view. Economics also plays some role in the selection bias. (While 50 cents is not a lot of money anymore, poor people are much more likely to think twice before spending the money to express their opinion.)

    (b) This survey was based on fairly standard modern-day polling techniques (random sample telephone interviews, etc.) but it had one subtle flaw. How reliable can a survey about the conduct of the newsmedia be when the survey itself is conducted by a newsmedia organization? ("The fox guarding the chicken-coop" syndrome.)

    (c) Both surveys seem to have produced unreliable data — survey 1 overestimating the public's dissapproval of the role played by the newsmedia and survey 2 overestimating the public's support for the press coverage of the war. In survey 1, only those with strong opinions were surveyed (i.e. called the 900 number). In survey 2, people without strong opinions were called. Further, they may have been swayed by being surveyed by a media organization.

    (d) Any reasoned out answer should be acceptable. (Since Area Code 900 telephone polls are particularly unreliable, survey 2 gets our vote.)

71. (a) Both samples should be a representative cross section of the same population. In particular, it is essential that the first sample, after being released, be allowed to disperse evenly throughout the population, and that the population should not change between the time of the capture and the time of the recapture.

    (b) It is possible (especially when dealing with elusive types of animals) that the very fact that the animals in the first sample allowed themselves to be captured makes such a sample biased (they could represent a slower, less cunning group). This type of bias is compounded with the animals that get captured the second time around. A second problem is the effect that the first capture can have on the captured animals. Sometimes the animal may be hurt (physically or emotionally) making it more (or less) likely to be captured the second time around. A third source of bias is the possibility that some of the tags will come off.

RUNNING

73. 625; we solve the two equations $pN = 250$ and $p(1-p)N = 150$ for p and N. But substituting $N = 250/p$ into the second equation gives $(1-p)250 = 150$ which yields $p = 0.4$. Thus, $N = 250/0.4 = 625$.

Chapter 15

WALKING

15.1 Graphs and Charts

1. (a)

Score	10	40	50	60	70	80	100
Frequency	1	1	3	7	6	5	2

(b) For each score, plot a bar with height represented by the frequency of that score in the data set.

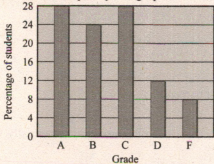

3. (a)

Grade	A	B	C	D	F
Frequency	7	6	7	3	2

(b) A relative frequency bar graph shows the *percentage* of data in each category.

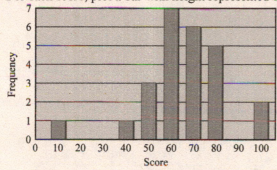

5. (a)

Distance	0.0	0.5	1.0	1.5	2.0	2.5	3.0	5.0	6.0	8.5
Frequency	4	4	3	7	3	2	1	1	1	1

(b)

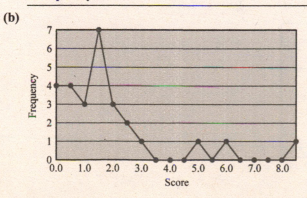

7.

Class Interval	Very close	Close	Nearby	Not too far	Far
Frequency	8	10	5	1	3

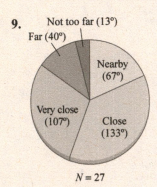

9.

$N = 27$

Slice "Very close": $\dfrac{8}{27} \times 360° \approx 106.7°$;

Slice "Close": $\dfrac{10}{27} \times 360° \approx 133.3°$;

Slice "Nearby": $\dfrac{5}{27} \times 360° \approx 66.7°$;

Slice "Not too far": $\dfrac{1}{27} \times 360° \approx 13.3°$;

Slice "Far": $\dfrac{3}{27} \times 360° = 40°$.

11. (a) $N = 4 + 6 + 7 + 5 + 6 + 8 + 2 + 2 = 40$

(b) 0%

(c) $\dfrac{5 + 6 + 8 + 2 + 2}{40} = 0.575 \approx 58\%$ (rounded to the nearest percent)

13. (a) qualitative; it is a variable that describes characteristics that cannot be measured numerically.

(b) $0.47 \times 19{,}548 \approx 9188$ died due to an accident.

15.

Class Interval	200-290	300-390	400-490	500-590	600-690	700-800
Relative Frequency	3.85%	15.55%	31.80%	29.20%	15.14%	4.45%

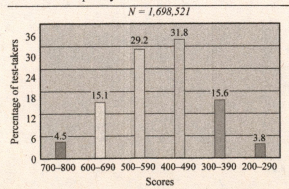

$N = 1{,}698{,}521$

17. (a)

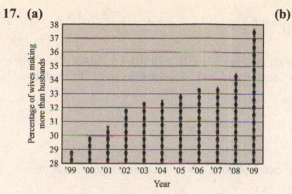

(b)

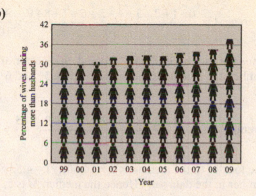

19. (a) 60 – 48 = 12 ounces

(b) The third class interval: "more than 72 ounces and less than or equal to 84 ounces." Values that fall exactly on the boundary between two class intervals belong to the class interval to the left.

(c)

Frequency	15	24	41	67	119	184	142	26	5	2
Percent	2.4	3.8	6.6	10.7	19.0	29.4	22.7	4.2	0.8	0.3

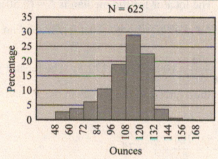

N = 625

21. (a) 9 + 6 + 2 = 17 (use Figure 15-17(a))

(b) 4 + 10 = 14 (use Figure 15-17(b))

(c) 14 – 13 = 1 (use Figures 15-17(a) and 15-17(b))

15.2 Means, Medians, and Percentiles

23. (a) Average $A = \dfrac{3 - 5 + 7 + 4 + 8 + 2 + 8 - 3 - 6}{9} = 2$

(b) The ordered data set is {-6, -5, -3, 2, 3, 4, 7, 8, 8}. The locator of the 50th percentile is $L = (0.50)(9) = 4.5$. Since L is not a whole number, the 50th percentile is located in the 5th position in the list. Hence, the median M is 3.

(c) The average of the new data set is most easily found by computing $\dfrac{(9 \times 2) + 2}{10} = 2$ The new ordered data set is {-6, -5, -3, 2, 2, 3, 4, 7, 8, 8}. The locator of the 50th percentile is $L = (0.50)(10) = 5$. Since L is a whole number, the 50th percentile is the average of the 5th and 6th numbers (2 and 3) in the ordered list. So, the median of this new data set is 2.5.

25. (a) Average $A = \dfrac{0+1+2+3+4+5+6+7+8+9}{10} = \dfrac{45}{10} = 4.5$

Since the data set is already ordered, the locator of the median is $L = (0.50)(10) = 5$. Since L is a whole number, the median is the average of the 5th and 6th numbers (4 and 5) in the data set. The median M is 4.5.

(b) Average $A = \dfrac{1+2+3+4+5+6+7+8+9}{9} = \dfrac{45}{9} = 5$

The locator of the median is $L = (0.50)(9) = 4.5$. Since L is not a whole number, the median is the 5th number in the data set. Hence the median M is 5.

(c) Average $A = \dfrac{1+2+3+4+5+6+7+8+9+10}{10} = \dfrac{55}{10} = 5.5$

The locator of the median is $L = (0.50)(10) = 5$. Since L is a whole number, the median is the average of the 5th and 6th numbers (5 and 6) in the data set. The median M is 5.5.

27. (a) Average $A = \dfrac{5+10+15+20+25+60}{6} = \dfrac{135}{6} = 22.5$. The locator for the median is $L = (0.50)(6) = 3$.
Hence, the median is the average of the values in the 3rd and 4th positions. That is, the median M is 17.5.

(b) Average $A = \dfrac{105+110+115+120+125+160}{6} = \dfrac{600+135}{6} = 100 + 22.5 = 122.5$. The locator for the
median is $L = (0.50)(6) = 3$. Once again the median is the average of the values in the 3rd and 4th positions. That is, the median M is 117.5.

29. (a) Average $A = \dfrac{24 \times 0 + 16 \times 1 + 20 \times 2 + 12 \times 3 + 5 \times 4 + 3 \times 5}{24 + 16 + 20 + 12 + 5 + 3} = \dfrac{127}{80} = 1.5875$

(b) The locator for the median is $L = (0.50)(80) = 40$. So, the median is the average of the 40th and 41st scores. The 40th score is 1, and the 41st score is 2. Thus, the median M is 1.5.

31. (a) Since the number of quiz scores N is not given, we can compute the average as
$$A = \dfrac{(0.07N) \times 4 + (0.11N) \times 5 + (0.19N) \times 6 + (0.24N) \times 7 + (0.39N) \times 8}{N}$$
$$= 0.07 \times 4 + 0.11 \times 5 + 0.19 \times 6 + 0.24 \times 7 + 0.39 \times 8 = 6.77$$

(As the calculation shows, the number of scores is not important. If one likes, they can assume that there were 100 scores.)

(b) Since 50% of the scores were at 7 or below, the median M is 7.

33. The ordered data set is {-6, -5, -3, 2, 4, 7, 8, 8}.

(a) The locator of the 25th percentile is $L = (0.25)(8) = 2$. Since this is a whole number, the 25th percentile is the average of the 2nd and 3rd values in the list (namely -5 and -3). That is, the 25th percentile is -4 ($Q_1 = -4$).

(b) The locator of the 75th percentile is $L = (0.75)(8) = 6$. Since this is a whole number, the 75th percentile is the average of the 6th and 7th values in the list (namely 7 and 8). That is, the 75th percentile is 7.5 ($Q_3 = 7.5$).

(c) The new ordered data set is {-6, -5, -3, 2, 2, 4, 7, 8, 8}. The locator of the 25th percentile is $L = (0.25)(9)$ $= 2.25$. Rounding up, we find that the first quartile is the 3rd value in the ordered list. That is, $Q_1 = -3$. The locator of the 75th percentile is $L = (0.75)(9) = 6.75$. Rounding up, we find that the third quartile is the 7th value in the ordered list. That is, $Q_3 = 7$.

35. (a) Since the data set is already ordered, the locator of the 75th percentile is given by $L = (0.75)(100) = 75$. Since this is a whole number, the 75th percentile is the average of the 75th and 76th values in the list. That is, the 75th percentile is 75.5.
The locator of the 90th percentile is given by $L = (0.90)(100) = 90$. Since this is a whole number, the 90th percentile is the average of the 90th and 91st values in the list. That is, the 90th percentile is 90.5.

(b) The locator of the 75th percentile is given by $L = (0.75)(101) = 75.75$. Since this not is a whole number, the 75th percentile is the 76th value in the list. That is, the 75th percentile is 75.

The locator of the 90th percentile is given by $L = (0.90)(101) = 90.9$. Since this is not a whole number, the 90th percentile is the 91st value in the list. That is, the 90th percentile is 90.

(c) The locator of the 75th percentile is given by $L = (0.75)(99) = 74.25$. Since this not is a whole number, the 75th percentile is the 75th value in the list. That is, the 75th percentile is 75.

The locator of the 90th percentile is given by $L = (0.90)(99) = 89.1$. Since this is not a whole number, the 90th percentile is the 90th value in the list. That is, the 90th percentile is 90.

(d) The locator of the 75th percentile is given by $L = (0.75)(98) = 73.5$. Since this not is a whole number, the 75th percentile is the 74th value in the list. That is, the 75th percentile is 74.

The locator of the 90th percentile is given by $L = (0.90)(98) = 88.2$. Since this is not a whole number, the 90th percentile is the 89th value in the list. That is, the 90th percentile is 89.

37. (a) The Cleansburg Fire Department consists of $2 + 7 + 6 + 9 + 15 + 12 + 9 + 9 + 6 + 4 = 79$ firemen. The locator of the first quartile is thus given by $L = (0.25)(79) = 19.75$. So, the first quartile is the 20th number in the ordered data set. That is, $Q_1 = 29$.

(b) The locator of the third quartile is given by $L = (0.75)(79) = 59.25$. So, the third quartile is the 60th number in the ordered data set. That is, $Q_3 = 32$.

(c) The locator of the 90th percentile is given by $L = (0.90)(79) = 71.1$. So, the 90th percentile is the 72nd number in the ordered data set or 37.

39. (a) The 836,198th score, $d_{836,198}$. [The locator is given by $L = 0.50(1,672,395) = 836,197.5$.]

(b) The 418,099st score, $d_{418,099}$. [The locator is given by $L = 0.25(1,672,395) = 418,098.75$.]

(c) The average of the 1,337,916th and 1,337,917th scores, $d_{1,337,916.5}$. [The locator is given by $L = 0.80(1,672,395) = 1,337,916$.]

41. (a) $Min = -6$, $Q_1 = -4$, $M = 3$, $Q_3 = 7.5$, $Max = 8$

(b)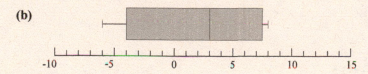

43. (a) $Min = 25, Q_1 = 29, M = 31, Q_3 = 32, Max = 39$

(b)

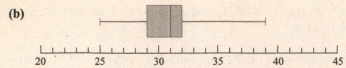

45. (a) \$43,000 (corresponding the vertical line in the middle of the lower box)

(b) \$50,000

(c) The vertical line indicating the median salary in the engineering box plot is aligned with the right end of the box in the agriculture box plot.

15.3 Ranges and Standard Deviations

47. (a) $8 - (-6) = 14$

(b) From Exercise 33, $Q_1 = -4, Q_3 = 7.5.$ $IQR = 7.5 - (-4) = 11.5.$

49. (a) \$156,000 - \$115,000 = \$41,000

(b) At least 171 homes

51. (a) Note that $1.5 \times IQR = 1.5 \times 3 = 4.5$. Any number bigger than $Q_3 + 1.5 \times IQR = 12 + 4.5 = 16.5$ is an outlier.

(b) Any number smaller than $Q_1 - 1.5 \times IQR = 9 - 4.5 = 4.5$ is also an outlier.

(c) Since 1 is the only value smaller than 4.5 and 24 is the only value larger than 16.5, the values 1 and 24 are the only outliers in this data set.

53. The *IQR* for 18-year old U.S. males is $71 - 67 = 4$ inches. Since $1.5 \times IQR = 1.5 \times 4 = 6$, an outlier is any height more than $Q_3 + 1.5 \times IQR = 71 + 6 = 77$ inches or any height less than $67 - 6 = 61$ inches.

55. (a) $A = 5$, so $x - A = 0$ for every number x in the data set. The standard deviation is 0.

(b) $A = \dfrac{0 + 5 + 5 + 10}{4} = 5$

x	$x - 5$	$(x-5)^2$
0	−5	25
5	0	0
5	0	0
10	5	25
		50

Standard deviation $= \sqrt{\dfrac{50}{4}} = \dfrac{5\sqrt{2}}{2} \approx 3.5$

(c) $A = \dfrac{0+10+10+20}{4} = 10$

x	$x-10$	$(x-10)^2$
0	-10	100
10	0	0
10	0	0
20	10	100
		200

Standard deviation $= \sqrt{\dfrac{200}{4}} = 5\sqrt{2} \approx 7.1$

(d) $A = \dfrac{1+2+3+4+5}{5} = 3$

x	$x-3$	$(x-3)^2$
1	-2	4
2	-1	1
3	0	0
4	1	1
5	2	4
		10

Standard deviation $= \sqrt{\dfrac{10}{5}} = \sqrt{2} \approx 1.4$

JOGGING

57. (a) 30

  (b) 8

59. Let x = score Mike needs on the next exam.

$\dfrac{5 \cdot 88 + x}{6} = 90$

$5 \cdot 88 + x = 540$

$x = 100$

61. Since the median is the average of $d_{732,872}$ and $d_{732,873}$, the locator of the median is $L = 732,872$. Since $(0.5)(N) = L = 732,872$, we find $N = 1,465,744$.

63. (a) {1, 1, 1, 1, 6, 6, 6, 6, 6, 6} Average = 4;
Median = 6.

(b) {1, 1, 1, 1, 1, 1, 6, 6, 6, 6} Average = 3;
Median =1.

(c) {1, 1, 6, 6, 6, 6, 6, 6, 6, 6} Average = 5;
$Q_1 = 6$.

(d) {1, 1, 1, 1, 1, 1, 1, 1, 6, 6} Average = 2;
$Q_3 = 1$.

65. From histogram (a) we deduce that the median team salary is between $70 and $100 million. From histogram (b) we deduce that the median team salary is between $50 and $80 million. It follows that the median team salary must be between $70 and $80 million.

67. (a) $\dfrac{(x_1 + c) + (x_2 + c) + (x_3 + c) + \ldots + (x_N + c)}{N}$

$= \dfrac{(x_1 + x_2 + x_3 + \ldots + x_N) + cN}{N}$

$= \dfrac{(x_1 + x_2 + x_3 + \ldots + x_N)}{N} + \dfrac{cN}{N} = A + c$

(b) The relative sizes of the numbers are not changed by adding a constant c to every number. Thus (assuming the original data set is sorted) if the median of the data set is $M = x_k$, then the median of the new data set will be $x_k + c = M + c$. If the median of the original set is

$M = \dfrac{x_k + x_{k+1}}{2}$, then the median of the new data set will be

$\dfrac{(x_k + c) + (x_{k+1} + c)}{2} = \dfrac{x_k + x_{k+1}}{2} + \dfrac{2c}{2}$

$= M + c$. In either case, we see that the median of the new data set is $M + c$.

69. (a) 4

$\dfrac{\text{Column area over interval } 30-35}{\text{Column area over interval } 20-30} = \dfrac{5 \times h}{10 \times 1} = \dfrac{50\%}{25\%}$

and so $5h = 20$, $h = 4$.

(b) 0.4

$\dfrac{\text{Column area over interval } 35-45}{\text{Column area over interval } 20-30} = \dfrac{10 \times h}{10 \times 1} = \dfrac{10\%}{25\%}$

and so $h = 0.4$.

(c) 0.4

$\dfrac{\text{Column area over interval } 45-60}{\text{Column area over interval } 20-30} = \dfrac{15 \times h}{10 \times 1} = \dfrac{15\%}{25\%}$

and so $h = 0.4$.

RUNNING

71. If the standard deviation is 0, then the average of the squared deviations is 0. The squared deviations cannot be negative, so if their average is 0, they must all be 0. It follows that all deviations from the average must be 0. This means that each data value is equal to the average a, i.e., the data set is constant.

73. $1 + 2 + ... + N = \dfrac{N(N+1)}{2}$ implies the average $A = \dfrac{1+2+...+N}{N} = \dfrac{N+1}{2}$. If N is odd, the median M is the "middle" number

$M = \dfrac{N+1}{2}$. If N is even the median M is the

average of $\dfrac{N}{2}$ and $\dfrac{N}{2}+1$, which is

$M = \dfrac{N+1}{2}$.

75. (a) 8/9 or approximately 89%

(b) 10

(c) Since $k > 0$, $1 - \dfrac{1}{k^2} < 1$ will always be the case.

Chapter 16

16.1 Sample Spaces and Events

1. **(a)** {*HHH, HHT, HTH, THH, TTH, THT, HTT, TTT*}
   Note that there are $2^3 = 8$ outcomes consisting of all possible sequences of *H*'s and *T*'s of length three.

   (b) {*sss, ssf, sfs, fss, ffs, fsf, sff, fff*}
   Note the correspondence between the outcomes in (a) and (b).

   (c) {0, 1, 2, 3}

   (d) {0, 1, 2, 3}

3. **(a)** $S = \{3, 4, 5, \ldots, 16, 17, 18\}$
   Each number rolled is an integer between 1 and 6. The minimum total that can be rolled is $1 + 1 + 1 = 3$. The maximum total that can be rolled is $6 + 6 + 6 = 18$. Every integer between 3 and 18 is possible.

   (b) $S = \{5, 3, 1, -1, -3, -5\}$
   If 0 heads are tossed (so that 5 tails are tossed), then $0 - 5 = -5$ is the observation. If 1 head is tossed (so that 4 tails are also tossed), the observation would be $1 - 4 = -3$. If 2 heads are tossed, the observation would be $2 - 3 = -1$. The following table summarizes how S is generated.

Heads Tossed	Tails Tossed	Observation
0	5	-5
1	4	-3
2	3	-1
3	2	1
4	1	3
5	0	5

5. **(a)** $S = \{AB, AC, AD, AE, BA, BC, BD, BE, CA, CB, CD, CE, DA, DB, DC, DE, EA, EB, EC, ED\}$
   Here, *XY* denotes the outcome that *X* was elected chairman and *Y* was elected treasurer. We have listed all possible permutations (here the order matters) of two members of the five-member board. Order clearly matters here as the outcome *BD* where *B* is elected chairman and *D* is elected treasurer is different from the outcome *DB* where *D* is elected chairman and *B* is elected treasurer. Note the manner in which we list the outcomes - it is organized alphabetically to ensure that all outcomes are indeed listed.

   (b) $S = \{ABC, ABD, ABE, ACD, ACE, ADE, BCD, BCE, BDE, CDE\}$
   We have listed all possible combinations (order does not matter here) of three for this five-member board. Here, for example, *ACE* denotes a search committee made up of *A*, *C*, and *E* (all committee members have equal roles so no order is needed in listing them). Note the manner in which we list the outcomes – it is once again organized alphabetically to ensure that all outcomes are listed.

7. Answers will vary. A typical outcome is a string of 10 letters each of which can be either an *H* or a *T*. An answer like {*HHHHHHHHHH*, …, *TTTTTTTTTT*} is not sufficiently descriptive. An answer like {… *HTTHHHTHTH*, …, *TTHTHHTTHT*, …, *HHHTHTTHHT*, …} is better. An answer like $\{X_1 X_2 X_3 X_4 X_5 X_6 X_7 X_8 X_9 X_{10} : \text{each } X_i \text{ is either } H \text{ or } T\}$ is best. Note: This sample space consists of $N = 2^{10} = 1024$ outcomes (2 possible outcomes at each of 10 stages of the experiment).

9. Answers will vary. An outcome is an ordered sequence of four numbers, each of which is an integer between 1 and 6 inclusive. The best answer would be something like $\{(n_1, n_2, n_3, n_4):$ each n_i is 1, 2, 3, 4, 5, or 6$\}$. An answer such as $\{(1,1,1,1), ..., (1,1,1,6), ...,(1,2,3,4),..., (3,2,6,2), ..., (6,6,6,6)\}$ showing a few typical outcomes is possible, but not as good. An answer like $\{(1,1,1,1),..., (2,2,2,2), ..., (6,6,6,6)\}$ is not descriptive enough. Note: This sample space consists of $N = 6^4 = 1296$ outcomes (6 possible outcomes at each of 4 stages of the experiment).

11. (a) $E_1 = \{HHT, HTH, THH\}$
       Remember that the event is all outcomes for which we observe two of the coins coming up heads.

    (b) $E_2 = \{HHH, TTT\}$

    (c) $E_3 = \{\}$
       There is no way to have half of three tosses come up heads.

    (d) $E_4 = \{TTH, TTT\}$

13. (a) $E_1 = \{(1,1),(2,2),(3,3),(4,4),(5,5),(6,6)\}$

    (b) $E_2 = \{(1,1),(1,2),(2,1),(6,6)\}$
       Note that outcome (1,2) and outcome (2,1) are different observations here. Both of these make up the event "a total of 3 is rolled."

    (c) $E_3 = \{(1,6),(2,5),(3,4),(4,3),(5,2),(6,1),(5,6),(6,5)\}$
       Again, outcomes (2,5) and (5,2) are different observations corresponding to a "2" being rolled on one die or the other.

15. (a) $E_1 = \{HHHHHHHHHH\}$

    (b) $E_2 = \{HHHHHHHHHT, HHHHHHHHTH, HHHHHHHTHH, HHHHHHTHHH, HHHHHTHHHH,$
       $HHHHTHHHHH, HHHTHHHHHH, HHTHHHHHHH, HTHHHHHHHH, THHHHHHHHH\}$
       Note that these outcomes are written in alphabetical order (an organized list).

    (c) $E_3 = \{TTTTTTTTTH, TTTTTTTTHT, TTTTTTTHTT, TTTTTTHTTT, TTTTTHTTTT,$
       $TTTTHTTTTT, TTTHTTTTTT, TTHTTTTTTT, THTTTTTTTT, HTTTTTTTTT, TTTTTTTTTT\}$
       Note that these outcomes are also written in alphabetical order. There are really two types of outcomes here – the 10 with exactly one *H* and the one with no *H*.

16.2 The Multiplication Rule, Permutations, and Combinations

17. (a) $9 \times 26^3 \times 10^3 = 158,184,000$
       Stage 1: There are 9 choices for the leading digit.
       Stages 2-4: There are 26 choices for the letter in slot 2, 26 choices for the letter in slot 3, and 26 choices for the letter in slot 3.
       Stages 5-7: There are 10 choices for the digit in slot 5, 10 choices for the digit in slot 6, and 10 choices for the digit in slot 7.

    (b) $1 \times 26^3 \times 10^2 \times 1 = 1,757,600$
       Stage 1: There is one choice for the leading digit (it must be a 5).
       Stages 2-4: There are 26 choices for the letter in slot 2, 26 choices for the letter in slot 3, and 26 choices for the letter in slot 3.
       Stages 5-6: There are 10 choices for the digit in slot 5 and 10 choices for the digit in slot 6.
       Stage 7: There is one choice for the last digit (it must be a 9).

(c) $9 \times 26 \times 25 \times 24 \times 9 \times 8 \times 7 = 70,761,600$

Stage 1: There are 9 choices for the leading digit.
Stages 2-4: There are 26 choices for the letter in slot 2, 25 choices for the letter in slot 3 (it must be different from the letter in slot 2), and 24 choices for the letter in slot 3 (it must be different than the letters in slots 2 and 3).
Stages 5-7: There are 9 choices for the digit in slot 5 (it must be different from the digit in slot 1), 8 choices for the digit in slot 6 (it must be different from the digits in slots 1 and 5), and 7 choices for the digit in slot 7 (it must be different from the digits in slots 1, 5, and 6).

19. (a) $(2+1) \times (3+4) \times 3 = 63$

Stage 1 (Footwear): Jack chooses from two pairs of shoes and a pair of boots (2 + 1 = 3 choices).
Stage 2 (Lower wear): Jack chooses from three pairs of jeans and four pairs of dress pants (3 + 4 = 7 choices).
Stage 3 (Upper wear): Jack chooses a dress shirt (3 options).

(b) $(2+1) \times (3+4) \times (3+3 \times 2) = 189$

Stage 1 (Footwear): Jack has 2 + 1 = 3 options
Stage 2 (Lower wear): Jack has 3 + 4 = 7 options.
Stage 3 (Upper wear): Jack chooses a dress shirt (3 options) or both a dress shirt and a jacket ($3 \times 2 = 6$ options). This totals $3 + 3 \times 2 = 9$ options.

21. (a) $8! = 40,320$

There are 8 choices for which book to place "first" (on the far left), 7 choices for which of the remaining books to place "second" (next to the first book), 6 choices for which of the remaining books to place "third", etc.

(b) $40,320 - 1 = 40,319$ (there is only one way in which all of the books are in order)

23. (a) $5 \times 4 \times 3 \times 2 \times 1 \times 4 \times 3 \times 2 \times 1 = 2880$

There are 5 choices for the first person (a woman) in line. Then, there are 4 choices for the second person (one of the remaining women) in line. Next, there are 3 choices for the third person (one of the three remaining women) in line. Etc. Once the number of ways to place the women at the front of the line have been determined (as 5!), the number of ways to place the men at the end of the line can be counted. This process is similar to that used in counting the number of ways to organize the women in line. There are 4 choices for the sixth person (a man) in line. Then, there are 3 choices for the seventh person (one of the remaining men) in line. Etc.

(b) $5 \times 4 \times 4 \times 3 \times 3 \times 2 \times 2 \times 1 \times 1 = 2880$

There are 5 choices for the first person (a woman) in line. Then, there are 4 choices for the second person (a man) in line. The third person in line is one of the 4 remaining women. The fourth person in line is one of the 3 remaining men. And so on.

25. (a) A typical outcome is a string of 10 letters each of which can be either an *H* or a *T*. This sample space consists of $N = 2^{10} = 1024$ outcomes (2 possible outcomes at each of 10 stages of the experiment).

(b) An outcome is an ordered sequence of four numbers, each of which is an integer between 1 and 6 inclusive. This sample space consists of $N = 6^4 = 1296$ outcomes (6 possible outcomes at each of 4 stages of the experiment).

(c) An outcome is sum of four numbers, each of which is an integer between 1 and 6 inclusive. This sample space consists of $N = 4 \times 6 - 4 \times 1 + 1 = 21$ outcomes (the sum must be between 4 and 24 inclusive).

27. (a) $_{15}P_4$; the order the candidates are selected to fill these openings matters (i.e., this is an ordered selection of 4 of the 15 board members).

(b) $_{15}C_4$; the order in which the committee is selected is irrelevant (i.e. this is an unordered selection of 4 of the 15 board members).

29. (a) $_{30}C_{18}$; the order the songs are selected for the set list does not matter (i.e., this an unordered selection of 18 of the 30 songs).

(b) $_{30}P_{18}$; the order the songs are selected for the CD (sequencing) matters (i.e. this is an ordered selection of 18 of the 30 songs).

31. (a) $_{40}C_{25}$; the order the players are listed on the active roster does not matter (i.e., this an unordered selection of 25 of the 40 players on the roster).

(b) $_{25}P_9$; the order the players are selected for the lineup matters (i.e. this is an ordered selection of 9 of the 25 active players).

16.3 Probabilities and Odds

33. (a) $\Pr(o_1) + \Pr(o_2) + \Pr(o_3) + \Pr(o_4) + \Pr(o_5) = 1$

$$0.22 + 0.24 + 3\Pr(o_3) = 1$$

$$\Pr(o_3) = 0.18$$

The probability assignment is $\Pr(o_1) = 0.22$, $\Pr(o_2) = 0.24$, $\Pr(o_3) = 0.18$, $\Pr(o_4) = 0.18$, $\Pr(o_5) = 0.18$.

(b) $\Pr(o_1) + \Pr(o_2) + \Pr(o_3) + \Pr(o_4) + \Pr(o_5) = 1$

$$0.22 + 0.24 + \left[\Pr(o_4) + 0.1\right] + \Pr(o_4) + 0.1 = 1$$

$$\Pr(o_4) = 0.17$$

The probability assignment is $\Pr(o_1) = 0.22$, $\Pr(o_2) = 0.24$, $\Pr(o_3) = 0.27$, $\Pr(o_4) = 0.17$, $\Pr(o_5) = 0.1$.

35. We know that $\Pr(A) = \dfrac{1}{5}$. If $\Pr(B) = x$, then $\Pr(C) = 2x$ and $\Pr(D) = 3x$. So, since these probabilities must sum to 1, we have $1/5 + x + 2x + 3x = 1$ which means that $6x = 4/5$ or $x = 2/15$. It follows that the probability assignment is $\Pr(A) = \dfrac{1}{5}, \Pr(B) = \dfrac{2}{15}, \Pr(C) = \dfrac{4}{15}, \Pr(D) = \dfrac{6}{15}$.

37. (a) $\Pr(E_1) = \dfrac{3}{8} = 0.375$; the sample space has 8 equally likely outcomes (see Exercise 1(a)). Of these outcomes, three of them constitute the event $E_1 = \{HHT, HTH, THH\}$ (see Exercise 11(a)).

(b) $\Pr(E_2) = \dfrac{2}{8} = 0.25$; see Exercises 1(a) and 11(b).

(c) $\Pr(E_3) = 0$; see Exercises 1(a) and 11(c).

(d) $\Pr(E_4) = \dfrac{2}{8} = 0.25$; see Exercises 1(a) and 11(d).

39. (a) $\Pr(E_1) = \dfrac{6}{36} = \dfrac{1}{6}$; the sample space has $6^2 = 36$ equally likely outcomes. Of these outcomes, six of them constitute the event $E_1 = \{(1,1), (2,2), (3,3), (4,4), (5,5), (6,6)\}$ (see Exercise 13(a)).

(b) $\Pr(E_2) = \dfrac{4}{36} = \dfrac{1}{9}$ (see Exercise 13(b))

(c) $\Pr(E_3) = \dfrac{8}{36} = \dfrac{2}{9}$ (see Exercise 13(c))

41. (a) $\Pr(E_1) = \dfrac{1}{1024} \approx 0.001$; the sample space has $2^{10} = 1024$ equally likely outcomes (see Exercise 7). Of these outcomes, only one of them constitutes the event that none of the tosses come up tails: $E_1 = \{HHHHHHHHHH\}$ (see Exercise 15(a)).

(b) $\Pr(E_2) = \dfrac{10}{1024} = \dfrac{5}{512} \approx 0.01$; (see Exercise 15(b))

(c) $\Pr(E_3) = \dfrac{11}{1024} \approx 0.01$; (see Exercise 15(c))

43. The total number of outcomes in this random experiment is $2^{10} = 1024$.

(a) There is only one way to get all ten correct. So, $\Pr(\text{getting 10 points}) = \dfrac{1}{1024}$. If it helps, think of this as being similar to Exercise 41(a). If the "key" to the quiz is all "True" (or "Heads"), there is only one outcome in which the student earns exactly 10 points.

(b) There is only one way to get all ten incorrect (and hence a score of –5). So,
$\Pr(\text{getting -5 points}) = \dfrac{1}{1024}$. Again, you can think of this as the "key" to the quiz consisting of all
"True" answers. In this case, there is only one outcome in which the student earns a score of -5 (their quiz would have all "False" answers. See Exercise 41 and use True=Heads and False=Tails.

(c) In order to get 8.5 points, the student must get exactly 9 correct answers and 1 incorrect answer. There are ${}_{10}C_1 = 10$ ways to select which answer would be answered incorrectly. So,
$\Pr(\text{getting 8.5 points}) = \dfrac{10}{1024} = \dfrac{5}{512}$. See also Exercise 41(b).

(d) In order to get 8 or more points, the student must get at least 9 correct answers (if they only get 8 correct answers, they lose a point for guessing two incorrect answers and score 7 points). So,
$\Pr(\text{getting 8 or more points}) = \Pr(\text{getting 8.5 points}) + \Pr(\text{10 points}) = \dfrac{10}{1024} + \dfrac{1}{1024} = \dfrac{11}{1024}$. See also Exercise 41(c).

(e) If the student gets 6 answers correct, they score $6 - 4 \times 0.5 = 4$ points. If the student gets 7 answers correct, they score $7 - 3 \times 0.5 = 5.5$ points. So, there is no chance of getting exactly 5 points. Hence, $\Pr(\text{getting 5 points}) = 0$.

(f) If the student gets 8 answers correct, they score $8 - 2 \times 0.5 = 7$ points. There are ${}_{10}C_2 = 45$ ways to select which 8 answers would be answered correctly. So, $\Pr(\text{getting 7 points}) = \dfrac{45}{1024}$. In order to get 7 or more points, the student needs to answer at least 8 answers correctly. $\Pr(\text{getting 7 or more points}) =$
$\Pr(\text{getting 7 points}) + \Pr(\text{getting 8.5 points}) + \Pr(\text{10 points}) = \dfrac{45}{1024} + \dfrac{10}{1024} + \dfrac{1}{1024} = \dfrac{56}{1024} = \dfrac{7}{128}$.

45. (a) There are $_{15}C_4 = 1365$ ways to choose four delegates. If Alice is selected, there are $_{14}C_3 = 364$ ways to choose the other three delegates. So, Pr(Alice selected) $= \dfrac{364}{1365} = \dfrac{4}{15}$.

   (b) Pr(Alice is not selected) $= 1 - \dfrac{364}{1365} = \dfrac{1001}{1365} = \dfrac{11}{15}$.

   (c) There are $_{15}C_4 = 1,365$ ways to select four members, but only one way to select Alice, Bert, Cathy, and Dale. Pr(Alice, Bert, Cathy, and Dale selected) $= \dfrac{1}{_{15}C_4} = \dfrac{1}{1,365}$.

47. The complementary event F to $E =$ "a tail comes up at least once" is $F =$ "a tail does not come up." The probability of this complementary event F is Pr(F) $= 1/2^{10} = 1/1024$. It follows that Pr(E) $= 1 -$ Pr(F) $= 1 - 1/1024 = 1023/1024$.

49. (a) $a = 4$, $b = 7$, $b - a = 7 - 4 = 3$. The odds in favor of E are 4 to 3. That is, for every 7 times the experiment is run, we should expect the event E to occur 4 times and to not occur 3 times.

   (b) $a = 6$, $b = 10$, $b - a = 10 - 6 = 4$. The odds in favor of E are 6 to 4, or 3 to 2.

51. (a) $\Pr(E) = \dfrac{3}{3+5} = \dfrac{3}{8}$

   (b) $\Pr(E) = 1 - \dfrac{8}{8+15} = 1 - \dfrac{8}{23} = \dfrac{15}{23}$

16.4 Expectations

53. Using the fact that $90/120 = 75\%$ and $144/180 = 80\%$, Paul's score in the course is the weighted average $0.15 \times 77\% + 0.15 \times 83\% + 0.15 \times 91\% + 0.1 \times 75\% + 0.25 \times 87\% + 0.2 \times 80\% = 82.9\%$.

55. $100\% - 7\% - 22\% - 24\% - 23\% - 19\% = 5\%$ are 19 years old. So, the average age at Thomas Jefferson HS is the weighted average $0.07 \times 14 + 0.22 \times 15 + 0.24 \times 16 + 0.23 \times 17 + 0.19 \times 18 + 0.05 \times 19 = 16.4$.

57. $E = \dfrac{1}{2} \times \$1 + \dfrac{1}{4} \times \$5 + \dfrac{1}{8} \times \$10 + \dfrac{1}{10} \times \$20 + \dfrac{1}{40} \times \$100 = \$7.50$

59. $E = \dfrac{1}{8} \times 0 + \dfrac{3}{8} \times 1 + \dfrac{3}{8} \times 2 + \dfrac{1}{8} \times 3 = 1.5$ heads

16.5 Measuring Risk

61. $E = \dfrac{1}{6} \times \$18 + \dfrac{1}{18} \times \$54 + \dfrac{14}{18} \times \$(-9) = \$(-1)$

63. $E = \dfrac{1}{38} \times \$36 + \dfrac{37}{38} \times \$(-1) = \$ - \dfrac{1}{38} \approx \$ - 0.0263$. So, for every \$100 bet on a particular number (like '10'), you should expect to lose \$2.63. For every \$1,000,000 bet, you should expect to lose about \$26,316.

65. There is a 24% chance that Joe will gain $400 - 80 = 320$. There is a 76% chance that Joe will lose his \$80. So, $E = 0.24 \times \$320 + 0.76 \times \$(-80) = \$16$. Joe should take this risk since his expected payoff is greater than 0 (\$16). That is, if he made this transaction on thousands of plasma TVs, in the long run he could expect to save \$16 for each warranty he purchases.

67.	Payoff (to insurer)	P	$(P-500)$	$(P-1,500)$	$(P-4,000)$
	Probability	50%	35%	12%	3%

In order to make an average profit of $50 per policy, we solve
$E = 0.5 \times \$P + 0.35 \times \$(P-500) + 0.12 \times \$(P-1,500) + 0.03 \times \$(P-4,000) = \$50$ for P. Doing this gives $P = \$525$. That is, the insurance company should charge $525 per policy.

JOGGING

69. (a) $35 \times 34 \times 33 = {}_{35}P_3 = 39,270$

   (b) $15 \times 14 \times 13 = {}_{15}P_3 = 2730$.

   (c) The total number of all-female committees is $15 \times 14 \times 13 = {}_{15}P_3 = 2730$.
   The total number of all-male committees is $20 \times 19 \times 18 = {}_{20}P_3 = 6840$.
   So, there are $2730 + 6840 = 9570$ committees of the same gender.

   (d) The remaining $35 \times 34 \times 33 - (15 \times 14 \times 13 + 20 \times 19 \times 18) = 39,270 - (2730 + 6840) = 29,700$ committees are mixed.

71. (a) The event that R wins the match in game 5 can be described by {*AARRR, ARARR, ARRAR, RAARR, RARAR, RRAAR*}.

   (b) The event that R wins the match can be described by {*RRR, ARRR, RARR, RRAR, AARRR, ARARR, ARRAR, RAARR, RARAR, RRAAR*}.

   (c) The event that the match goes five games can be described by { *AARRR, ARARR, ARRAR, RAARR, RARAR, RRAAR, RRAAA, RARAA, RAARA, ARRAA, ARARA, AARRA*}.

73. (a) A sample space S for this random experiment might be taken to consist of all ways that 3 balls could be selected from the 10 in the urn (*without* regard to order). That is $N = {}_{10}C_3 = 120$. The number of outcomes in the event that two balls drawn are blue and one is red is ${}_3C_2 \times {}_7C_1 = 21$. So, Pr(two are blue and one is red) = 21/120 = 7/40.

   (b) In this random experiment, each draw is an independent event. So, we can multiply the probabilities of each event. Pr(only first and third balls are blue) = $\dfrac{3}{10} \times \dfrac{7}{10} \times \dfrac{3}{10} = \dfrac{63}{1000}$.

   (c) A sample space S for this random experiment might be taken to consist of all ways that 3 balls could be selected from the 10 in the urn (*with* regard to order). That is $N = {}_{10}P_3 = 720$. The number of outcomes in the event that only the first and third balls drawn are blue (making the second ball drawn red) is ${}_3P_2 \times {}_7P_1 = 42$. So, Pr(first and third balls are blue) = 42/720 = 7/120.

75. If N is odd, the probability is 0 (there can't be the same number of H's as T's). If N is even, say $N = 2K$, then the probability is ${}_NC_K \left(\dfrac{1}{2}\right)^N = \dfrac{N!}{2^N K!K!} = \dfrac{(2K)!}{2^N K!K!}$.

77. The key word in this exercise is "at least." The sample space consists of all possible outcomes. There are $_{500}C_5$ ways in which 5 tickets can be selected from the 500 to be "winning" tickets. For the event that you win at least one prize to occur, it could happen that you win one prize, two prizes, or three prizes. So, in symbolic notation,

    $\text{Pr(winning at least one prize)} = \text{Pr(win 1 prize)} + \text{Pr(win 2 prizes)} + \text{Pr(win 3 prizes)}$

    $$= \frac{_5C_1 \cdot {}_{495}C_2}{_{500}C_5} + \frac{_5C_2 \cdot {}_{495}C_1}{_{500}C_5} + \frac{_5C_3 \cdot {}_{495}C_0}{_{500}C_5}$$

79. **(a)** $\text{Pr(YAHTZEE)} = \dfrac{6}{6^5} = \dfrac{6}{7776} \approx 0.00077$

    The denominator, 7776, is the number of ordered outcomes possible. The numerator consists of the only 6 outcomes that give YAHTZEE (six 1's, six 2's, etc...).

    (b) $\text{Pr(four of a kind)} = \dfrac{5 \times 6 \times 5}{6^5} = \dfrac{150}{7776} \approx 0.019$

    There are 5 choices as to which die will not match the other 4. There are 6 choices for what will constitute the four of kind (four sixes, four fives, etc...) and there are 5 choices for what will constitute the other die.

    (c) $\text{Pr(large straight)} = \dfrac{2 \times 5!}{6^5} = \dfrac{240}{7776} \approx 0.031$

    There are 2 choices as to what the large straight will be – it will either be 1-5 or 2-6. For each there are 5! ways to rearrange the order that the dice are rolled.

Chapter 17

WALKING

17.2 Normal Curves and Normal Distributions

1. (a) $\mu = 80$ lb; it is located in the exact middle of the distribution.

    (b) $M = 80$ lb; it too is located in the exact middle of the distribution.

    (c) $\sigma = 90$ lb. $- 80$ lb. $= 10$ lb; it is the horizontal distance between the middle of the distribution and the inflection point P.

3. (a) $\dfrac{32 \text{ in.} + 48 \text{ in.}}{2} = 40$ in.

    (b) Using the result from (a), we see that $\sigma = 40$ in. $- 32$ in. $= 8$ in.

    (c) $Q_3 = 40$ in. $+ 0.675 \times 8$ in. $= 45.4$ in.

    (d) $Q_1 = 40$ in. $- 0.675 \times 8$ in. $= 34.6$ in.

5. (a) $Q_3 \approx 81.2$ lb. $+ 0.675 \times 12.4$ lb. ≈ 89.6 lb

    (b) $Q_1 \approx 81.2$ lb. $- 0.675 \times 12.4$ lb. ≈ 72.8 lb

7. (a) $\mu = \dfrac{432.5 \text{ points} + 567.5 \text{ points}}{2}$
    $= 500$ points

    (b) $567.5 \approx 500 + 0.675 \cdot \sigma$
    $\sigma \approx 100$ points

9. $94.7 \approx 81.2 + 0.675 \times \sigma$
    $\sigma \approx 20$ in.

11. $\mu \neq M$; in any normal distribution these two values must be the same.

13. In a normal distribution the first and third quartiles are the same distance from the mean. In this distribution, $\mu - Q_1 = 195 - 180 = 15$ and $Q_3 - \mu = 220 - 195 = 25$.

15. (a) $\dfrac{45 \text{ kg} - 30 \text{ kg}}{15 \text{ kg}} = \dfrac{15 \text{ kg}}{15 \text{ kg}} = 1$

    (b) $\dfrac{0 \text{ kg} - 30 \text{ kg}}{15 \text{ kg}} = \dfrac{-30 \text{ kg}}{15 \text{ kg}} = -2$

    (c) $\dfrac{54 \text{ kg} - 30 \text{ kg}}{15 \text{ kg}} = \dfrac{24 \text{ kg}}{15 \text{ kg}} = 1.6$

    (d) $\dfrac{3 \text{ kg} - 30 \text{ kg}}{15 \text{ kg}} = \dfrac{-27 \text{ kg}}{15 \text{ kg}} = -1.8$

17. (a) In a normal distribution, the third quartile is about 0.675 standard deviations above the mean. That is, $Q_3 \approx \mu + (0.675)\sigma$. In this case, it means $391 \approx 310 + (0.675)\sigma$ so that the standard deviation is $\sigma \approx 120$. Hence, the standardized value of 490 points is approximately $\dfrac{490 - 310}{120} = 1.5$.

    (b) $\dfrac{250 - 310}{120} = -0.5$

    (c) $\dfrac{220 - 310}{120} = -0.75$

    (d) $\dfrac{442 - 310}{120} = 1.1$

19. (a) 183.5 feet
    $\dfrac{x - 183.5}{31.2} = 0$
    $x - 183.5 = 0$
    $x = 183.5$ ft

    (b) 152.3 feet
    $\dfrac{x - 183.5}{31.2} = -1$
    $x - 183.5 = -31.2$
    $x = 152.3$ ft

21. (a) 199.1 feet
    $\dfrac{x - 183.5}{31.2} = 0.5$
    $x - 183.5 = 15.6$
    $x = 199.1$ ft

(b) 111.74 feet

$$\frac{x-183.5}{31.2}=-2.3$$

$$x-183.5=-71.76$$

$$x=111.74 \text{ ft}$$

23. $\dfrac{84-50}{\sigma}=2$

$$34=2\sigma$$

$$\sigma=17 \text{ lb.}$$

25. $\dfrac{50-\mu}{15}=3$

$$50-\mu=45$$

$$\mu=5$$

27. $\dfrac{20-\mu}{\sigma}=-2;\dfrac{100-\mu}{\sigma}=3$

$$20-\mu=-2\sigma;100-\mu=3\sigma$$

$$\mu=20+2\sigma, \text{ so } 100-(20+2\sigma)=3\sigma$$

$$100-20-2\sigma=3\sigma$$

$$80=5\sigma$$

$$\sigma=16$$

$$\mu=20+2\times16=52$$

29. 54 inches is one standard deviation above the mean, and 42 inches is one standard deviation below the mean.

$$\mu=\frac{42 \text{ in.}+54 \text{ in.}}{2}=48 \text{ in.}$$

$$\sigma=54 \text{ in.}-48 \text{ in.}=6 \text{ in.}$$

31. (a) 98.8 cm is two standard deviations above the mean and 85.2 cm is two standard deviations below the mean. So,

$$\mu=\frac{85.2 \text{ cm}+98.8 \text{ cm}}{2}=92 \text{ cm}$$

$$\sigma=\frac{1}{2}(98.8 \text{ cm}-92 \text{ cm})=3.4 \text{ cm}$$

(b) $Q_1=92 \text{ cm}-0.675\times3.4 \text{ cm}\approx89.7 \text{ cm}$

$$Q_3=92 \text{ cm}+0.675\times3.4 \text{ cm}=94.3 \text{ cm}$$

33. (a) The 16th percentile is one standard deviation below the 50th percentile. So,

$$\sigma=71.5 \text{ in.}-65.2 \text{ in.}=6.3 \text{ in.}$$

(b) The 84th percentile is one standard deviation above the 50th percentile. So,

$$P_{84}=71.5 \text{ in.}+6.3 \text{ in.}=77.8 \text{ in.}$$

35. Since 97.5% of the data lies below two standard deviations above the mean,

$$P_{97.5}=\mu+2\sigma=\mu+2(6.1 \text{ cm})=81.5 \text{ cm}.$$

Thus, $\mu=81.5 \text{ cm}-2(6.1 \text{ cm})=69.3 \text{ cm}$.

37. 9.9 has a standardized value of

$\dfrac{9.9-12.6}{4}=-0.675$. Also, 16.6 has a

standardized value of $\dfrac{16.6-12.6}{4}=1$. So,

25% of the data is below the first quartile of 9.9. Also, 84% of the data is below 16.6. So, 84% - 25% = 59% of the data is between 9.9 and 16.6.

39. (a) 95% of the data lies within two standard deviations of the mean. Hence, 2.5% of the data are not within two standard deviations on each side of the mean. So, 97.5% of the data fall below $\mu+2\sigma$, the point two standard deviations above the mean.

(b) 97.5% of the data lies below $z=2$ (i.e. below $\mu+2\sigma$). Since 68% of the data lies within one standard deviation of the mean, 32% of the data is not within one standard deviation of the mean. So, 68% + 16% = 84% of the data fall below $z=1$ (i.e. below $\mu+\sigma$). It follows that 97.5% – 84% = 13.5% of the data falls between $z=1$ and $z=2$ (i.e. between $\mu+\sigma$ and $\mu+2\sigma$).

17.3 Modeling Approximately Normal Distributions

41. (a) 52 points

(b) 50%

(c) $\dfrac{41-52}{11}=-1,\dfrac{63-52}{11}=1$

So, since about 68% of data fall within one standard deviation of the mean (between standardized scores of -1 and 1) in a normal distribution, we would estimate that 68% of the students would score between 41 and 63 points.

(d) $\frac{1}{2}(100\% - 68\%) = 16\%$

Or, since 50% + 34% = 84% of the students score 63 points or less, approximately 16% of the students score 63 points or more.

43. (a) $Q_1 \approx 52 - 0.675 \times 11 \approx 44.6$ points

(b) $Q_3 \approx 52 + 0.675 \times 11 \approx 59.4$ points

45. (a) $\frac{99-125}{13} = -2, \frac{151-125}{13} = 2$

95% have blood pressure between 99 and 151 mm (i.e., between standardized scores of -2 and 2). 95% of 2000 patients is 1900 patients ($0.95 \times 2000 = 1900$).

(b) 112 is one standard deviation below the mean. 151 is two standard deviations above the mean. The percentage of patients falling between one standard deviation below the mean and two standard deviations above the mean is 68% + 13.5% = 81.5%. 81.5% of the 2000 patients is 1630 patients.

47. (a) Since $\frac{112 \text{ mm} - 125 \text{ mm}}{13 \text{ mm}} = -1, x = 112$
mm is at the 16th percentile (one standard deviation below the mean).

(b) Since $\frac{138 \text{ mm} - 125 \text{ mm}}{13 \text{ mm}} = 1, x = 138$ mm
is at the 84th percentile (one standard deviation above the mean).

(c) Since $\frac{164 \text{ mm} - 125 \text{ mm}}{13 \text{ mm}} = 3, x = 164$
mm is at the 99.85th percentile (three standard deviations above the mean).

49. (a) Since $\frac{11 \text{ oz} - 12 \text{ oz}}{0.5 \text{ oz}} = -2$ and

$\frac{13 \text{ oz} - 12 \text{ oz}}{0.5 \text{ oz}} = 2$, the rule of 95 tells us
that there is approximately a 95% chance that the bag weighs somewhere between 11 and 13 ounces.

(b) Since $\frac{13 \text{ oz} - 12 \text{ oz}}{0.5 \text{ oz}} = 2$, the rule of 95
tells us that there is approximately a 47.5% chance that the bag weighs

somewhere between 12 and 13 ounces (this is half of 95%).

(c) Because the chance of the bag weighing between 11 and 12 ounces is the same as the chance that it weighs between 12 and 13 ounces, we can use our answer to part (b) and symmetry to obtain an answer of 47.5% + 50% = 97.5%.

51. (a) Since $\frac{11 \text{ oz} - 12 \text{ oz}}{0.5 \text{ oz}} = -2$ the rule of 95
tells us that there is approximately a
$\frac{1}{2}(100\% - 95\%) = 2.5\%$ chance that any
one bag weighs less than 11 ounces. But, $0.025 \times 500 = 12.5 \approx 13$ bags.

(b) Since $\frac{11.5 \text{ oz} - 12 \text{ oz}}{0.5 \text{ oz}} = -1$ the rule of 68
tells us that there is approximately a
$\frac{1}{2}(100\% - 68\%) = 16\%$ chance that any
one bag weighs less than 11.5 ounces. But, $0.16 \times 500 = 80$ bags.

(c) $0.50 \times 500 = 250$ bags

(d) Since $\frac{12.5 \text{ oz} - 12 \text{ oz}}{0.5 \text{ oz}} = 1$ the rule of 68
tells us that there is approximately a
$\frac{1}{2}(68\%) + 50\% = 84\%$ chance that any
one bag weighs less than 12.5 ounces. But, $0.84 \times 500 = 420$ bags.

(e) Since $\frac{13 \text{ oz} - 12 \text{ oz}}{0.5 \text{ oz}} = 2$ the rule of 95
tells us that there is approximately a
$\frac{1}{2}(95\%) + 50\% = 97.5\%$ chance that any
one bag weighs less than 13 ounces. But, $0.975 \times 500 = 487.5 \approx 488$ bags.

(f) Since $\frac{13.5 \text{ oz} - 12 \text{ oz}}{0.5 \text{ oz}} = 3$ the rule of 99.7
tells us that there is approximately a
$\frac{1}{2}(99.7\%) + 50\% = 99.85\%$ chance that
any one bag weighs less than 13.5 ounces. But, $0.9985 \times 500 = 499.25 \approx 499$ bags.

53. (a) Since $\dfrac{7.21\ \text{kg} - 8.16\ \text{kg}}{0.95\ \text{kg}} = -1$, $x = 7.21\ \text{kg}$

is at the 16th percentile (one standard deviation below the mean).

(b) Since $\dfrac{10\ \text{kg} - 8.16\ \text{kg}}{0.95\ \text{kg}} \approx 2$, $x = 10\ \text{kg}$ is at

the 97.5th percentile (two standard deviations above the mean).

(c) The 75th percentile corresponds to the third quartile. This is a value located 0.675 of a standard deviation above the mean. This weight is thus $x =$ $8.16\ \text{kg} + 0.675 \times 0.95\ \text{kg} \approx 8.8\ \text{kg}$.

55. (a) Since $\dfrac{5.3\ \text{kg} - 4.5\ \text{kg}}{0.4\ \text{kg}} = 2$, $x = 5.3\ \text{kg}$ is at

the 97.5th percentile (two standard deviations above the mean).

(b) Since $\dfrac{5.7\ \text{kg} - 4.5\ \text{kg}}{0.4\ \text{kg}} = 3$, $x = 5.7\ \text{kg}$ is at

the 99.85th percentile (three standard deviations above the mean).

(c) The 25th percentile corresponds to 0.675 standard deviations below the mean. Her weight is thus $x =$ $4.5\ \text{kg} - 0.675 \times 0.4\ \text{kg}$ $\approx 4.23\ \text{kg}$.

17.4. Normality in Random Events

57. (a) $\mu = \dfrac{3600}{2} = 1800, \sigma = \dfrac{\sqrt{3600}}{2} = 30$

(b) Since $\dfrac{1770 - 1800}{30} = -1$ and

$\dfrac{1830 - 1800}{30} = 1$ there is a 68% chance

that Y will fall in this range (between one standard deviation below the mean and one standard deviation above the mean).

(c) The chance of this happening is

$\dfrac{1}{2}(68\%) = 34\%$ since this range is

between the mean and one standard deviation above the mean. That is, between $z = 0$ and $z = 1$.

(d) 1830 corresponds to a z-value of $z = 1$ and 1860 corresponds to a z-value of $z = 2$

(since $\dfrac{1860 - 1800}{30} = 2$). So,

$\dfrac{1}{2}(95\%) = 47.5\%$ of the number of tails

tossed should fall between $z = 0$ and $z = 2$ (i.e. between 1800 and 1860). $47.5\% - 34\% = 13.5\%$ of the number of tails tossed should fall between $z = 1$ and $z = 2$ (i.e. between 1830 and 1860).

59. $\mu = \dfrac{7056}{2} = 3528, \sigma = \dfrac{\sqrt{7056}}{2} = 42$

(a) Since $\dfrac{3486 - 3528}{42} = -1$ and

$\dfrac{3570 - 3528}{42} = 1$ there is a 68% chance

that the number of females will fall in this range (between one standard deviation below the mean and one standard deviation above the mean).

(b) $\dfrac{1}{2}(100\% - 68\%) = 16\%$

(c) $\dfrac{1}{2}(68\%) + 50\% = 84\%$

61. (a) $\mu = 600 \times 0.4 = 240,$

$\sigma = \sqrt{600 \times 0.4 \times (1 - 0.4)} = 12$

(b) $Q_1 \approx 240 - 0.675 \times 12 \approx 232$
$Q_3 \approx 240 + 0.675 \times 12 \approx 248$

(c) Since $\dfrac{216 - 240}{12} = -2$ and $\dfrac{264 - 240}{12} = 2$

there is a 95% chance that the number of coins will fall in this range (between two standard deviations below the mean and two standard deviations above the mean).

63. (a) Since $p = \dfrac{1}{6}$, $\mu = 180 \times \dfrac{1}{6} = 30$, and

$\sigma = \sqrt{180 \times \dfrac{1}{6} \times \left(1 - \dfrac{1}{6}\right)} = 5$.

(b) Since $\frac{40-30}{5}=2$, $x=40$ is two standard deviations above the mean. The rule of 95 gives a 2.5% chance of rolling a '6' more than 40 times.

(c) Since $\frac{35-30}{5}=1$, $x=35$ is one standard deviation above the mean. The rule of 68 gives a 34% (half of 68%) chance of rolling a '6' somewhere between 30 and 35 times.

JOGGING

65. 0

In a normal distribution, the mean and the median are the same values. Since the z-value of the mean is 0, so too is the z-value for the median.

67. (a) His weight is approximately
$8.16\text{ kg}+1.65\times0.95\text{ kg}=9.73\text{ kg}$.

(b) His weight is approximately
$8.16\text{ kg}-0.84\times0.95\text{ kg}=7.36\text{ kg}$.

69. (a) Since $\frac{8.4\text{ kg}-8.16\text{ kg}}{0.95\text{ kg}}\approx0.253$, $x=8.4$ kg is at approximately the 60th percentile (i.e. at $\mu+0.25\sigma$).

(b) Since $\frac{7.67\text{ kg}-8.16\text{ kg}}{0.95\text{ kg}}\approx-0.52$, $x=7\frac{2}{3}$ kg is at approximately the 30th percentile (i.e. at $\mu-0.52\sigma$).

71. (a) A score at the 75th percentile (third quartile) would be approximately 0.675 standard deviations above the mean. This score would be $511+(0.675)(120)=592$ points. Since SAT scores are always multiples of 10, we estimate that a score of 590 or 600 points would be at the third quartile.

(b) A score at the 84th percentile would be approximately one standard deviation above the mean using the 68-95-99.7 rule. This score would be $511+(1)(120)=631$ points. Since SAT scores are always multiples of 10, we estimate that a score of 630 or 640 points would be at the 84th percentile.

(c) A score at the 70th percentile would be approximately 0.52 standard deviations above the mean using Table 17-9. This score would be $511+(0.52)(120)=573.4$ points. Since SAT scores are always multiples of 10, we estimate that a score of 570 or 580 points would be at the 70th percentile.

73. (a) The 90th percentile of the data is located at $\mu+1.28\times\sigma=65.2+1.28\times10=78$ points.

(b) The 70th percentile of the data is located at $\mu+0.52\times\sigma=65.2+0.52\times10=70.4$ points.

(c) The 30th percentile of the data is located at $\mu-0.52\times\sigma=65.2-0.52\times10=60$ points.

(d) The 5th percentile of the data is located at $\mu-1.65\times\sigma=65.2-1.65\times10=48.7$ points.

RUNNING

75. The mean number of tails tossed is $\mu=\frac{n}{2}$. There is a 16% chance Y is more than one standard deviation above the mean, $\frac{n}{2}+\sigma$. Using $\sigma=\frac{\sqrt{n}}{2}$ where $\sigma=10$, we solve $10=\frac{\sqrt{n}}{2}$ for n. But $20=\sqrt{n}$ gives $n=400$.

77. $n=2500$
X has an approximately normal distribution with mean $\mu=\frac{n}{10}$ and standard deviation $\sigma=\sqrt{n\cdot\frac{1}{10}\cdot\frac{9}{10}}=\frac{3}{10}\sqrt{n}$. If there is a 95% chance that X will be between $\frac{n}{10}-30$ and $\frac{n}{10}+30$, then $30=2\sigma$ and $\sigma=15$. Solving $\frac{3}{10}\sqrt{n}=15$ gives $n=2500$.

79. (a) The probability of losing a bet on red is $20/38 \approx 0.53$. By the dishonest coin principle, Y has an approximately normal distribution with center $\mu \approx 10,000 \times 0.53$ $= 5300$ and standard deviation $\sigma \approx \sqrt{10000 \times 0.53 \times 0.47} \approx 50$.

(b) Since the center of the distribution is at $Y = 5300$, the chances that we will lose 5300 times or more ($Y \geq 5300$) are approximately 50%.

(c) The chances that we will lose somewhere between 5150 and 5450 times (i.e., Y is within plus or minus 3 standard deviations of the center) are 99.7%.

(d) To break even or win we must have $Y \leq 5000$ (i.e., Y must be more than 4 standard deviations left of the center). The chances of this are essentially 0.